Honda NB, ND, NP & NS50 Melody Owners Workshop Manual

by Chris Rogers

with an additional Chapter on the NB, ND and NP50 models

by Jeremy Churchill

Models covered
NB50 MS-E Melody deluxe. 49cc. March 1984 to April 1985
ND50 M-C Melody II deluxe. 49cc. April 1982 to April 1985
NP50 D Melody mini. 49cc. February 1983 to April 1985
NS50 D-B Melody 49cc. June 1981 to February 1984
NS50 MS-B Melody deluxe. 49cc. June 1981 to February 1984

Note: the NT50 Melody Mini models sold from April 1985 onwards are not covered in this manual

ISBN 978 1 85010 292 2

(622-3U5)

J H Haynes & Co. Ltd.
Haynes North America, Inc

www.haynes.com

British Library Cataloguing in Publication Data

Rogers, Chris
Honda NB, ND, NP & NS50 Melody owners workshop manual. —(Owners workshop manual).
1. Honda Melody motorcycle
I. Title II. Churchill, Jeremy
629.28'775 JL448.H6
ISBN 978-1 85010-292-2

Acknowledgements

Special thanks are due to Paul and Chris Branson of Paul Branson Motorcycles Limited of Yeovil, for their invaluable advice, the supply of service information and for providing the Honda Melody deluxe featured in this Manual.

The Avon Rubber Company kindly supplied information and technical assistance on tyre fitting and NGK Spark Plugs (UK) Limited supplied information on spark plug maintenance and electrode conditions.

About this manual

The purpose of this manual is to present the owner with a concise and graphic guide which will enable him to tackle any operation from basic routine maintenance to a major overhaul. It has been assumed that any work would be undertaken without the luxury of a well-equipped workshop and a range of manufacturer's service tools.

To this end, the machine featured in the manual was stripped and rebuilt in our own workshop, by a team comprising a mechanic, a photographer and the author. The resulting photographic sequence depicts events as they took place, the hands shown being those of the author and the mechanic.

The use of specialised, and expensive, service tools was avoided unless their use was considered to be essential due to risk of breakage or injury. There is usually some way of improvising a method of removing a stubborn component, providing that a suitable degree of care is exercised.

The author learnt his motorcycle mechanics over a number of years, faced with the same difficulties and using similar facilities to those encountered by most owners. It is hoped that this practical experience can be passed on through the pages of this manual.

Where possible, a well-used example of the machine is chosen for the workshop project, as this highlights any areas which might be particularly prone to giving rise to problems. In this way, any such difficulties are encountered and resolved before the text is written, and the techniques used to deal with them can be incorporated in the relevant section. Armed with a working knowledge of the machine, the author undertakes a considerable amount of research in order that the maximum amount of data can be included in the manual.

A comprehensive section, preceding the main part of the manual, describes procedures for carrying out the routine maintenance of the machine at intervals of time and mileage. This section is included particularly for those owners who wish to ensure the efficient day-to-day running of their motorcycle, but who choose not to undertake overhaul or renovation work.

Each Chapter is divided into numbered sections. Within these sections are numbered paragraphs. Cross reference throughout the manual is quite straightforward and logical. When reference is made 'See Section 6.10' it means Section 6, paragraph 10 in the same Chapter. If another Chapter were intended, the reference would read, for example, 'See Chapter 2, Section 6.10'. All the photographs are captioned with a section/paragraph number to which they refer and are relevant to the Chapter text adjacent.

Figures (usually line illustrations) appear in a logical but numerical order, within a given Chapter. Fig. 1.1 therefore refers to the first figure in Chapter 1.

Left-hand and right-hand descriptions of the machines and their components refer to the left and right of a given machine when the rider is seated normally.

Motorcycle manufacturers continually make changes to specifications and recommendations, and these, when notified, are incorporated into our manuals at the earliest opportunity.

We take great pride in the accuracy of information given in this manual, but motorcycle manufacturers make alterations and design changes during the production run of a particular motorcycle of which they do not inform us. No liability can be accepted by the authors or publishers for loss, damage or injury caused by any errors in, or omissions from, the information given.

Contents

The 1982 Honda Melody deluxe

Engine/transmission unit of the Honda Melody deluxe

Introduction to the Honda Melody

The present Honda empire, which started in a wooden shack in 1947, now occupies vast factory space, covering all aspects of modern motorcycle design, research, testing and production. The facilities available to the large work force are second to none.

The first motorcycle to be imported into the UK in the early 1960s, the 250 cc twin cylinder 'Dream', was the thin edge of the wedge which has been the Japanese domination of the motorcycle industry. Strange it looked too, to Western eyes, with pressed steel frame, and 'square' styling.

There was, however, nothing strange about the way these modern machines were accepted, and sales were impressive. The trend set by these early machines, that of almost unheard of sophistication, especially on machines of small engine capacity, allied to a remarkably reasonable selling price and a lively performance, ensured their success and started the name Honda on its way to becoming known in virtually every home in the country.

Honda now offer machines to the public to cover every conceivable aspect of motorcycling. These include prestigious sports machines of transverse four and six-cylinder engine configurations, a specialized flat four engine tourer, four and two-stroke engine moto-cross machines, a plush 500 cc V-twin engine turbo-charged sports machine, and even a three-wheeled go-anywhere, off-road fun machine.

The Melody represents Honda's first serious attempt to exploit a previously untapped market which has already been opened up by at least one of their major Japanese competitors. The recent dramatic increases in world oil prices have made many car and public transport users think seriously about alternative methods of getting to and from work. In addition there are many women who would not consider a powered two-wheeler as a practical means of transport, being wary of the motorcycle's masculine and oily image. To attract these potential buyers, Honda have opted for the already proven format and have produced an extremely economical fully automatic moped, with nearly all of the mechanical components hidden behind body panels. The result is a blend of Japanese technology and Italian scooter styling and layout. The number of controls has been kept to a minimum, often at the expense of astonishing complexity beneath the body panels, a good example being the sophisticated vacuum fuel tap and choke control valve combination. Both of these components are fully automatic in operation and have no manual controls but rely instead on a complex network of pipelines to effect their operation.

The handlebar controls of the Melody are limited to the necessary electrical switches, two brake levers and a throttle control. The electric starter motor incorporated in all Melody deluxe models and the fitting of fully automatic transmission, means that riding the machine is made as simple as possible, and is simplified to the stage where the rider needs only to press the starter button, open the throttle to accelerate away and apply the brakes to stop.

Equipped as standard with a combination of front and rear mounted carriers and with a comprehensive range of Hondastyle accessories to choose from, the Honda Melody has, at the time of writing this Manual, already made a good impression on its intended market.

Model dimensions and weight

Overall length	1520 mm (59.8 in)
Overall width	630 mm (24.8 in)
Overall height	960 mm (37.8 in)
Ground clearance	105 mm (4.13 in)
Wheelbase	1090 mm (42.9 in)
Dry weight	53 kg (116.9 lb)

Ordering spare parts

When replacement parts are required for your Honda, it is advisable to deal direct with a recognised Honda agent or with the area distributor. They are better placed to supply the parts ex-stock and should have the technical experience that may not be available with other suppliers. When ordering parts, always quote the engine and frame numbers in full, since these will identify the model and its date of manufacture. It will sometimes help if the old part is presented when the replacement is ordered to aid correct identification.

Always fit replacement parts of Honda manufacture and do not be tempted to use pattern parts. Although the pattern parts may appear similar they often give inferior service and may prove more expensive in the long run. In addition the use of parts not supplied by Honda may invalidate subsequent warranty claims.

The engine number of the machine is stamped on the right-hand face of the transmission casing and can be viewed through the rear wheel. The frame number is included on the information plate which is attached to the forward facing side of the frame downtube. This number is also stamped on a frame bracket which is adjacent to the mounting for the HT coil.

Some of the more expendable parts such as spark plugs, bulbs, tyres, oil and greases etc can be obtained from accessory shops and motor factors, who have convenient opening hours, and can often be found not far from home. It is also possible to obtain parts on a Mail Order basis from a number of specialists who advertise regularly in the motorcycle magazines.

Location of frame number

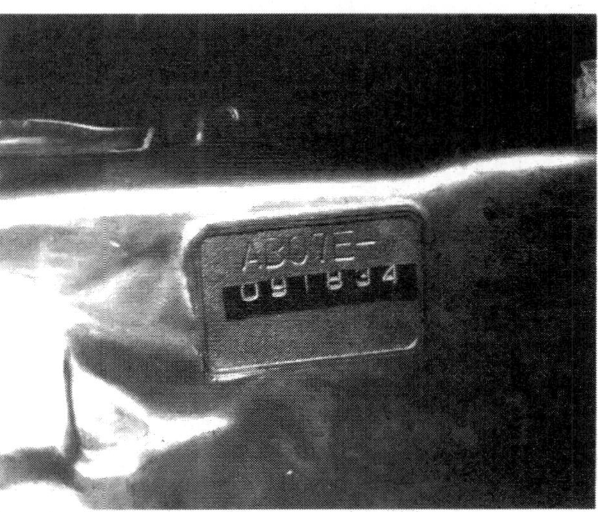

Location of engine number

Safety first!

Professional motor mechanics are trained in safe working procedures. However enthusiastic you may be about getting on with the job in hand, do take the time to ensure that your safety is not put at risk. A moment's lack of attention can result in an accident, as can failure to observe certain elementary precautions.

There will always be new ways of having accidents, and the following points do not pretend to be a comprehensive list of all dangers; they are intended rather to make you aware of the risks and to encourage a safety-conscious approach to all work you carry out on your vehicle.

Essential DOs and DON'Ts

DON'T start the engine without first ascertaining that the transmission is in neutral.

DON'T suddenly remove the filler cap from a hot cooling system – cover it with a cloth and release the pressure gradually first, or you may get scalded by escaping coolant.

DON'T attempt to drain oil until you are sure it has cooled sufficiently to avoid scalding you.

DON'T grasp any part of the engine, exhaust or silencer without first ascertaining that it is sufficiently cool to avoid burning you.

DON'T allow brake fluid or antifreeze to contact the machine's paintwork or plastic components.

DON'T syphon toxic liquids such as fuel, brake fluid or antifreeze by mouth, or allow them to remain on your skin.

DON'T inhale dust – it may be injurious to health (see *Asbestos* heading).

DON'T allow any spilt oil or grease to remain on the floor – wipe it up straight away, before someone slips on it.

DON'T use ill-fitting spanners or other tools which may slip and cause injury.

DON'T attempt to lift a heavy component which may be beyond your capability – get assistance.

DON'T rush to finish a job, or take unverified short cuts.

DON'T allow children or animals in or around an unattended vehicle.

DON'T inflate a tyre to a pressure above the recommended maximum. Apart from overstressing the carcase and wheel rim, in extreme cases the tyre may blow off forcibly.

DO ensure that the machine is supported securely at all times. This is especially important when the machine is blocked up to aid wheel or fork removal.

DO take care when attempting to slacken a stubborn nut or bolt. It is generally better to pull on a spanner, rather than push, so that if slippage occurs you fall away from the machine rather than on to it.

DO wear eye protection when using power tools such as drill, sander, bench grinder etc.

DO use a barrier cream on your hands prior to undertaking dirty jobs – it will protect your skin from infection as well as making the dirt easier to remove afterwards; but make sure your hands aren't left slippery. Note that long-term contact with used engine oil can be a health hazard.

DO keep loose clothing (cuffs, tie etc) and long hair well out of the way of moving mechanical parts.

DO remove rings, wristwatch etc, before working on the vehicle – especially the electrical system.

DO keep your work area tidy – it is only too easy to fall over articles left lying around.

DO exercise caution when compressing springs for removal or installation. Ensure that the tension is applied and released in a controlled manner, using suitable tools which preclude the possibility of the spring escaping violently.

DO ensure that any lifting tackle used has a safe working load rating adequate for the job.

DO get someone to check periodically that all is well, when working alone on the vehicle.

DO carry out work in a logical sequence and check that everything is correctly assembled and tightened afterwards.

DO remember that your vehicle's safety affects that of yourself and others. If in doubt on any point, get specialist advice.

IF, in spite of following these precautions, you are unfortunate enough to injure yourself, seek medical attention as soon as possible.

Asbestos

Certain friction, insulating, sealing, and other products – such as brake linings, clutch linings, gaskets, etc – contain asbestos. *Extreme care must be taken to avoid inhalation of dust from such products since it is hazardous to health.* If in doubt, assume that they *do* contain asbestos.

Fire

Remember at all times that petrol (gasoline) is highly flammable. Never smoke, or have any kind of naked flame around, when working on the vehicle. But the risk does not end there – a spark caused by an electrical short-circuit, by two metal surfaces contacting each other, by careless use of tools, or even by static electricity built up in your body under certain conditions, can ignite petrol vapour, which in a confined space is highly explosive.

Always disconnect the battery earth (ground) terminal before working on any part of the fuel or electrical system, and never risk spilling fuel on to a hot engine or exhaust.

It is recommended that a fire extinguisher of a type suitable for fuel and electrical fires is kept handy in the garage or workplace at all times. Never try to extinguish a fuel or electrical fire with water.

Note: *Any reference to a 'torch' appearing in this manual should always be taken to mean a hand-held battery-operated electric lamp or flashlight. It does* **not** *mean a welding/gas torch or blowlamp.*

Fumes

Certain fumes are highly toxic and can quickly cause unconsciousness and even death if inhaled to any extent. Petrol (gasoline) vapour comes into this category, as do the vapours from certain solvents such as trichloroethylene. Any draining or pouring of such volatile fluids should be done in a well ventilated area.

When using cleaning fluids and solvents, read the instructions carefully. Never use materials from unmarked containers – they may give off poisonous vapours.

Never run the engine of a motor vehicle in an enclosed space such as a garage. Exhaust fumes contain carbon monoxide which is extremely poisonous; if you need to run the engine, always do so in the open air or at least have the rear of the vehicle outside the workplace.

The battery

Never cause a spark, or allow a naked light, near the vehicle's battery. It will normally be giving off a certain amount of hydrogen gas, which is highly explosive.

Always disconnect the battery earth (ground) terminal before working on the fuel or electrical systems.

If possible, loosen the filler plugs or cover when charging the battery from an external source. Do not charge at an excessive rate or the battery may burst.

Take care when topping up and when carrying the battery. The acid electrolyte, even when diluted, is very corrosive and should not be allowed to contact the eyes or skin.

If you ever need to prepare electrolyte yourself, always add the acid slowly to the water, and never the other way round. Protect against splashes by wearing rubber gloves and goggles.

Mains electricity and electrical equipment

When using an electric power tool, inspection light etc, always ensure that the appliance is correctly connected to its plug and that, where necessary, it is properly earthed (grounded). Do not use such appliances in damp conditions and, again, beware of creating a spark or applying excessive heat in the vicinity of fuel or fuel vapour. Also ensure that the appliances meet the relevant national safety standards.

Ignition HT voltage

A severe electric shock can result from touching certain parts of the ignition system, such as the HT leads, when the engine is running or being cranked, particularly if components are damp or the insulation is defective. Where an electronic ignition system is fitted, the HT voltage is much higher and could prove fatal.

Working conditions and tools

When a major overhaul is contemplated, it is important that a clean, well-lit working space is available, equipped with a workbench and vice, and with space for laying out or storing the dismantled assemblies in an orderly manner where they are unlikely to be disturbed. The use of a good workshop will give the satisfaction of work done in comfort and without haste, where there is little chance of the machine being dismantled and reassembled in anything other than clean surroundings. Unfortunately, these ideal working conditions are not always practicable and under these latter circumstances when improvisation is called for, extra care and time will be needed.

The other essential requirement is a comprehensive set of good quality tools. Quality is of prime importance since cheap tools will prove expensive in the long run if they slip or break when in use, causing personal injury or expensive damage to the component being worked on. A good quality tool will last a long time, and more than justify the cost.

For practically all tools, a tool factor is the best source since he will have a very comprehensive range compared with the average garage or accessory shop. Having said that, accessory shops often offer excellent quality tools at discount prices, so it pays to shop around. There are plenty of tools around at reasonable prices, but always aim to purchase items which meet the relevant national safety standards. If in doubt, seek the advice of the shop proprietor or manager before making a purchase.

The basis of any tool kit is a set of open-ended spanners, which can be used on almost any part of the machine to which there is reasonable access. A set of ring spanners makes a useful addition, since they can be used on nuts that are very tight or where access is restricted. Where the cost has to be kept within reasonable bounds, a compromise can be effected with a set of combination spanners – open-ended at one end and having a ring of the same size on the other end. Socket spanners may also be considered a good investment, a basic 3/8 in or 1/2 in drive kit comprising a ratchet handle and a small number of socket heads, if money is limited. Additional sockets can be purchased, as and when they are required. Provided they are slim in profile, sockets will reach nuts or bolts that are deeply recessed. When purchasing spanners of any kind, make sure the correct size standard is purchased. Almost all machines manufactured outside the UK and the USA have metric nuts and bolts, whilst those produced in Britain have BSF or BSW sizes. The standard used in USA is AF, which is also found on some of the later British machines. Others tools that should be included in the kit are a range of crosshead screwdrivers, a pair of pliers and a hammer.

When considering the purchase of tools, it should be remembered that by carrying out the work oneself, a large proportion of the normal repair cost, made up by labour charges, will be saved. The economy made on even a minor overhaul will go a long way towards the improvement of a toolkit.

In addition to the basic tool kit, certain additional tools can prove invaluable when they are close to hand, to help speed up a multitude of repetitive jobs. For example, an impact screwdriver will ease the removal of screws that have been tightened by a similar tool, during assembly, without a risk of damaging the screw heads. And, of course, it can be used again to retighten the screws, to ensure an oil or airtight seal results. Circlip pliers have their uses too, since gear pinions, shafts and similar components are frequently retained by circlips that are not too easily displaced by a screwdriver. There are two types of circlip pliers, one for internal and one for external circlips. They may also have straight or right-angled jaws.

One of the most useful of all tools is the torque wrench, a form of spanner that can be adjusted to slip when a measured amount of force is applied to any bolt or nut. Torque wrench settings are given in almost every modern workshop or service manual, where the extent to which a complex component, such as a cylinder head, can be tightened without fear of distortion or leakage. The tightening of bearing caps is yet another example. Overtightening will stretch or even break bolts, necessitating extra work to extract the broken portions.

As may be expected, the more sophisticated the machine, the greater is the number of tools likely to be required if it is to be kept in first class condition by the home mechanic. Unfortunately there are certain jobs which cannot be accomplished successfully without the correct equipment and although there is invariably a specialist who will undertake the work for a fee, the home mechanic will have to dig more deeply in his pocket for the purchase of similar equipment if he does not wish to employ the services of others. Here a word of caution is necessary, since some of these jobs are best left to the expert. Although an electrical multimeter of the AVO type will prove helpful in tracing electrical faults, in inexperienced hands it may irrevocably damage some of the electrical components if a test current is passed through them in the wrong direction. This can apply to the synchronisation of twin or multiple carburettors too, where a certain amount of expertise is needed when setting them up with vacuum gauges. These are, however, exceptions. Some instruments, such as a strobe lamp, are virtually essential when checking the timing of a machine powered by CDI ignition system. In short, do not purchase any of these special items unless you have the experience to use them correctly.

Although this manual shows how components can be removed and replaced without the use of special service tools (unless absolutely essential), it is worthwhile giving consideration to the purchase of the more commonly used tools if the machine is regarded as a long term purchase Whilst the alternative methods suggested will remove and replace parts without risk of damage, the use of the special tools recommended and sold by the manufacturer will invariably save time.

HONDA MELODY

Check list

Pre-riding checks

1 Check the amount of fuel indicated by the gauge is sufficient for your journey
2 Check that the oil level warning light is extinguished a few seconds after the ignition is switched on
3 Check the operation of the direction indicator lamps, indicator warning light, stop lamp and horn
4 Inspect the tyres for damage and low pressure
5 Check the operation of both the front and rear brakes
6 Start the machine and allow it to idle
7 Ensure that the headlamp, tail lamp and speedometer illumination light all function correctly
8 Check the throttle twistgrip operates smoothly and returns to idle speed when released
9 Check that the rear view mirror is correctly positioned and any luggage is properly secured

Weekly or every 20 miles (30 km)

1 Ensure that all electrical components function correctly and that the tyre tread complies with the legal requirements
2 Inspect the machine for loose nuts and bolts, frayed control cables and damaged tyres
3 Check the tyre pressures (cold)
4 Lubricate exposed portions of control cables

Monthly or every 80 miles (125 km)

1 Check the battery electrolyte level and top up if necessary

Annually or every 1000 miles (1500 km)

1 Inspect the fuel and vacuum pipes for leakage and deterioration
2 Remove and clean the fuel filter
3 Remove and clean the air filter element
4 Inspect the choke control filter for wear and clean the element
5 Check the throttle and oil pump cables for correct adjustment
6 Adjust the carburettor
7 Remove, clean and adjust the spark plug
8 Lubricate the control cables
9 Check the operation of the front and rear suspension
10 Examine the condition of the wheels and check for rim alignment
11 Adjust the front and rear brakes
12 Check the level of oil in the reduction gearbox and replenish if necessary

Every two years or 2000 miles (3000 km)

1 Change the reduction gearbox oil
2 Remove and examine the clutch shoes for wear
3 Decarbonise the engine and exhaust system

Additional service items

1 Cleaning the machine
2 Oil filter examination and cleaning

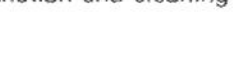

Adjustment data

Tyre pressures	
Front	21 psi (1.50 kg/cm²)
Rear	25 psi (1.75 kg/cm²)
Spark plug type	NGK BPR4HS or ND W14 FPR-L
Spark plug gap	0.6 – 0.7 mm (0.024 – 0.028 in)
Ignition timing	15° BTDC fixed
Engine idle speed	1800 ± 150 rpm

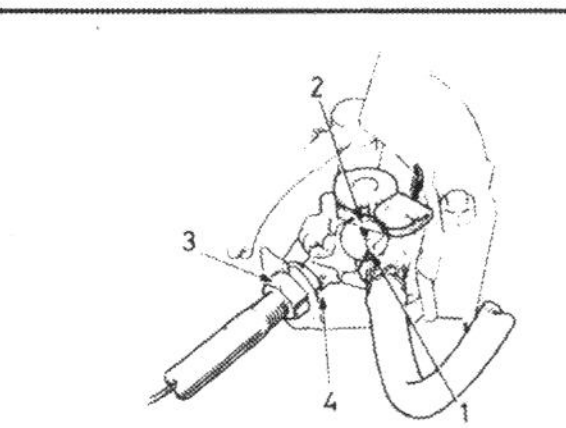

Oil pump cable adjustment synchronisation marks

1 Pump body index mark
2 Lever alignment mark
3 Adjusting nut
4 Locknut

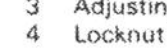

Recommended lubricants

Component	Quantity	Type/viscosity
1 Engine	1.2 lit (2.1 Imp pt)	Honda 2-stroke injector oil or equivalent
2 Reduction gearbox	90 cc (3.16 Imp fl oz)	SAE 10W/30 motor oil
3 Wheel bearings	As required	High melting point grease
4 Steering head bearings	As required	High melting point grease
5 Controls and pivots	As required	Medium weight general purpose grease
6 General lubrication and control cables	As required	Light machine oil

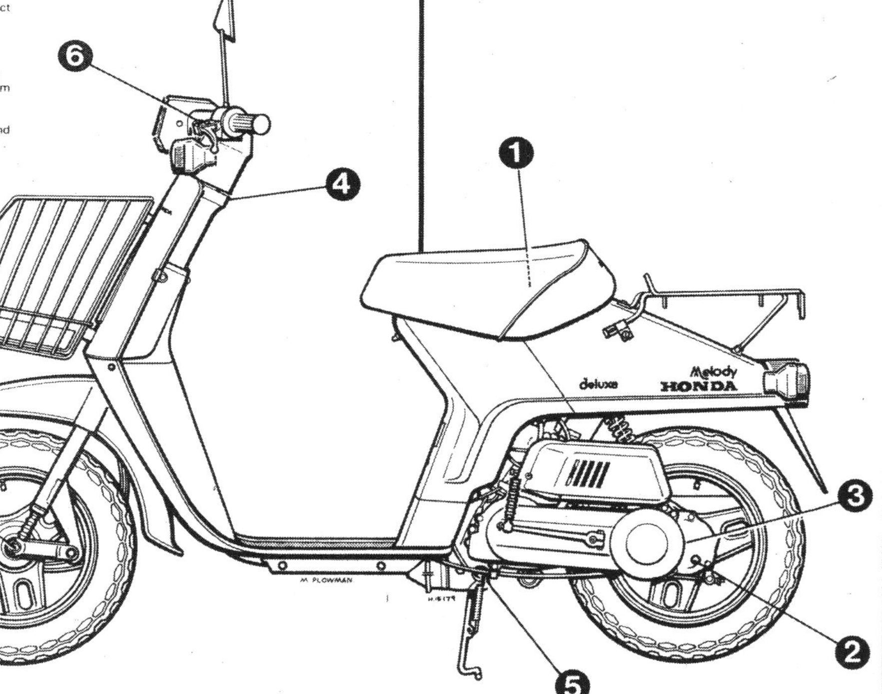

ROUTINE MAINTENANCE GUIDE

Routine maintenance

Periodic routine maintenance is a continuous process which should commence immediately the machine is used. The object is to maintain all adjustments and to diagnose and rectify minor defects before they develop into more extensive, and often more expensive, problems.

It follows that if the machine is maintained properly, it will both run and perform with optimum efficiency, and be less prone to unexpected breakdowns. Regular inspection of the machine will show up any parts which are wearing, and with a little experience, it is possible to obtain the maximum life from any one component, renewing it when it becomes so worn that it is liable to fail.

Regular cleaning can be considered as important as mechanical maintenance. This will ensure that all the cycle parts are inspected regularly and are kept free from accumulations of road dirt and grime.

Cleaning is especially important during the winter months, despite its appearance of being a thankless task which very soon seems pointless. On the contrary, it is during these months that the paintwork, chromium plating, and the alloy casings suffer the ravages of abrasive grit, rain and road salt. A couple of hours spent weekly on cleaning the machine will maintain its appearance and value, and highlight small points, like chipped paint, before they become a serious problem.

The various maintenance tasks are described under their respective mileage and calendar headings, and are accompanied by diagrams and photographs where pertinent.

It should be noted that the intervals between each maintenance task serve only as a guide. As the machine gets older, or if it is used under particularly arduous conditions, it is advisable to reduce the period between each check.

For ease of reference, most service operations are described in detail under the relevant heading. However, if further general information is required, this can be found under the pertinent Section heading and Chapter in the main text.

Although no special tools are required for routine maintenance, a good selection of general workshop tools is essential. Included in the tools must be a range of metric ring or combination spanners, a selection of crosshead screwdrivers. Additionally, owing to the extreme tightness of some component securing screws on the Honda Melody, an impact screwdriver, together with a suitable choice of cross and hexagon-headed bits, is absolutely indispensable. This is particularly so if the engine has not been dismantled since leaving the factory.

Pre-ride checks

The following quick checks should be undertaken prior to riding the machine. With a little practice they should become second nature and will only take a few minutes to carry out. Though simple, they will serve to alert the rider of any potentially dangerous or expensive faults.

1. Turn the ignition switch to its 'On' position and observe the position of the fuel level indicator needle
2. With the ignition on, the oil level warning light should illuminate for a few seconds then go out. If the light remains on, replenish the oil tank
3. Check the operation of the direction indicator lamps, the direction indicator warning light, the stop lamp and the horn
4. Check both tyres for obvious signs of damage or low pressure
5. Roll the machine off its centre stand and check both brakes for correct operation
6. Start the engine and allow it to idle. Remember that the starter will only work with the rear brake lever locked in the on position
7. Check the headlamp and tail lamp for correct operation. If riding in the dark, check the parking lamp and speedometer light
8. Check that the throttle twistgrip operates smoothly and returns to the idle position when released
9. Allow the engine to idle for a few minutes before moving off. Use this time to check that the rear view mirror is correctly adjusted and that any luggage is properly secured.

If any faults are noted, or if further information is required, refer to the following pages for detailed Routine Maintenance procedures.

Weekly, or every 20 miles (30 km)

1 Legal check

Check the operation of the electrical system, ensuring that the lights and horn are working properly and that the lenses are clean. Note that in the UK it is an offence to use a vehicle on which the lights are defective. This applies even when the machine is used during daylight hours. The horn is also a statutory requirement.

Inspect each tyre to ensure that the depth of its tread is above that required by law.

2 Safety check

Give the machine a close visual inspection, checking for loose nuts and fittings, frayed control cables etc. Check the tyres for damage, especially splitting on the sidewalls. Remove any stones or other objects caught between the treads. This is particularly important on the front tyre, where rapid deflation due to penetration of the inner tube will almost certainly cause total loss of control.

3 Tyre pressures

Check the tyre pressures. Always check with the tyres cold, using a pressure gauge known to be accurate. It is recommended that a pocket pressure gauge is purchased to offset any fluctuation between garage forecourt instruments. The correct tyre pressures are:-

Front	*21 psi (1.50 kg/cm²)*
Rear	*25 psi (1.75 kg/cm²)*

4 Control cable lubrication

Apply a few drops of oil to the exposed lengths of inner cable at the tops of the various control cables. This will prevent the cables drying out between the more thorough lubrication given during the annual/1000 mile service.

Monthly, or every 80 miles (125 km)

Carry out the service operations listed under the previous heading, then complete the following:

1 Battery check

Raise the seat of the machine. Release the battery retaining strap by unscrewing its single retaining screw and remove the translucent cover from the top of the battery. Disconnect each battery lead and carefully lift the battery from position whilst noting the location of its vent pipe in the battery housing. Place the battery on a flat and level surface.

The translucent plastic case of the battery permits the upper and lower levels of the electrolyte to be observed. Maintenance is normally limited to keeping the electrolyte level between the prescribed upper and lower limits and by making sure that the vent pipe is not blocked. The lead plates and their separators can be seen through the case, a further guide to the general condition of the battery.

Unless acid is spilt, as may occur if the machine falls over, the electrolyte should always be topped up with distilled water, to restore the correct level. If acid is spilt on any part of the machine, it should be neutralised with an alkali such as washing soda and washed away with plenty of water, otherwise serious corrosion will occur. Top up with sulphuric acid of the correct specific gravity (1.270 – 1.290) only when spillage has occurred.

When fitting the battery, ensure that the vent pipe is located correctly in the housing and remove any signs of corrosion from the terminal connections. Smear the terminals lightly with petroleum jelly (not grease) to prevent recurrence of the corrosion.

Annually, or every 1000 miles (1500 km)

Carry out the service operations listed under the previous headings, then complete the following:

1 Fuel and vacuum pipe examination

The condition of the fuel and vacuum pipes connected to the fuel tap should be checked at the above stated intervals. The synthetic rubber pipes are quite resistant to deterioration, but may eventually develop leaks where they push over their respective stubs. This can be corrected by disconnecting the pipe and slicing off the worn end. For obvious reasons this cannot be repeated many times before renewal becomes necessary.

Always obtain replacement pipes of the correct type, especially where identification marks are present. Replacement pipes of different materials should be avoided, and natural rubber tubing must be avoided at all costs. This latter material is attached by fuel and will disintegrate internally blocking the carburettor jets with a thick rubbery sludge.

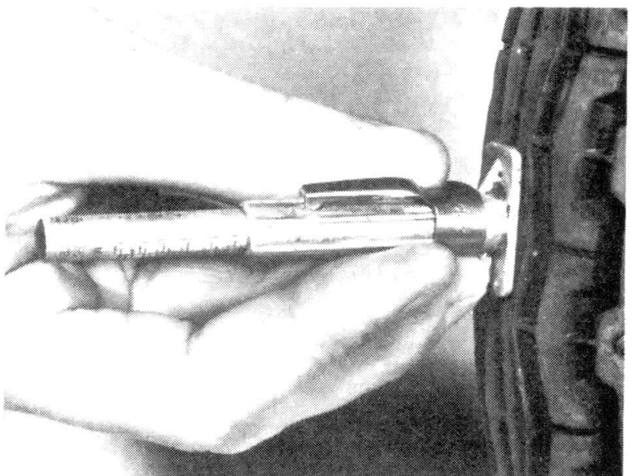

Check the depth of each tyre tread

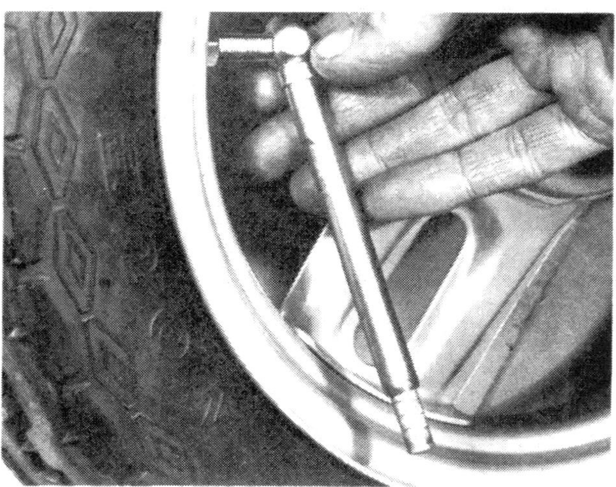

Check the pressure in each tyre

Lift the battery cover

Check the electrolyte level

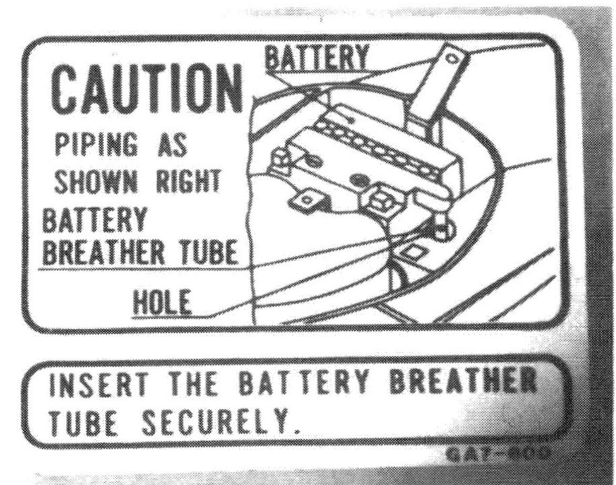
CAUTION
BATTERY
PIPING AS SHOWN RIGHT
BATTERY BREATHER TUBE
HOLE
INSERT THE BATTERY BREATHER TUBE SECURELY.
GA7-600

Route the battery vent correctly

Inspect all fuel and vacuum pipes

2 *Fuel filter cleaning*

The fuel filter is located over the stack pipe of the fuel tap. It is therefore necessary to remove the fuel tap from its location in the base of the fuel tank in order to gain access to the filter. Expose the tap by removing the centre body panel from the machine. This panel is retained in position by a single dome head nut with washer and a single screw. Before disconnecting any part of the tap assembly, take note of the following safety precautions. Take great care to place the machine in a well ventilated space where there is no chance of a build-up of fuel vapour occurring with the subsequent risk of a fire or explosion. Drain any fuel into a clean metal container that is equipped with a properly sealed lid; never use an open or plastic container. Store this container in a well ventilated open space. Never smoke or expose the machine to any kind of naked flame whilst draining fuel.

It is now necessary to completely drain the tank of fuel and then to remove the fuel tap from the tank. Draining of the tank can be achieved by either applying suction to the vacuum pipe in order to allow fuel to pass from the tank into a container or by removing the tank and inverting it to allow fuel to drain out of the fuel filler point.

With the tank drained of fuel and the fuel and vacuum pipes detached from the tap, unscrew the locknut which serves to retain the fuel tap in position and carefully pull the tap out of its location whilst taking care not to cause damage to the filler stack. Draw the small air filter off the end of the fuel tap air pipe and carefully remove the clip located beneath it. The filter can now be removed from the tap together with its base gasket.

Clean the fuel filter by removing it in clean fuel. Any stubborn traces of contamination can be removed by gently brushing the filter with a soft-bristled brush which has been soaked in fuel; a used toothbrush is ideal. Complete the cleaning process by blowing the filter clear with a jet of compressed air. Remember to take the necessary fire precautions whilst carrying out this cleaning procedure and always wear eye protection against any fuel that may spray back from the brush or, air jet. On completion of cleaning, closely inspect the filter for any splits or holes that will allow the passage of sediment through it and into the carburettor. Renew the filter if it is in any way defective.

Inspect and, if necessary, renew the small base gasket to the filter. Carefully clean the small air filter with a jet of compressed air and carry out a general inspection of the fuel tap for any signs of damage. Make sure that the air pipe is clear of contamination and fit the fuel filter with its base gasket over it. Push the fuel filter down into the centre of the tap locknut to seat it and then fit to the air pipe the support clip for the air filter. This clip must be located so that it is 17 mm (0.67 in) from the top of the air pipe. Fit the air filter and then carefully relocate the fuel tap in the tank. Hold the tap so that its fuel and vacuum pipe retaining stubs are correctly positioned and tighten the locknut. Pour a small amount of fuel into the tank and check for any leakage of fuel from around the tap to tank joint. If fuel is seen to leak from this joint and the leak cannot be stopped by further tightening of the nut, then it will be necessary to remove the tap and relocate or renew the fuel filter base gasket. Do not overtighten the tap locknut.

Finally, after having fitted the fuel tap and reconnected the fuel and vacuum pipes, refill the fuel tank and start the engine. Allow the engine to run at tick-over speed whilst carrying out a thorough check for leaks at all of the disturbed connection points. If a leak is found, it must be cured before the machine is ridden, otherwise there is a considerable risk of fire resulting in serious injury to the rider. On completion of a satisfactory leak check, refit the centre body panel.

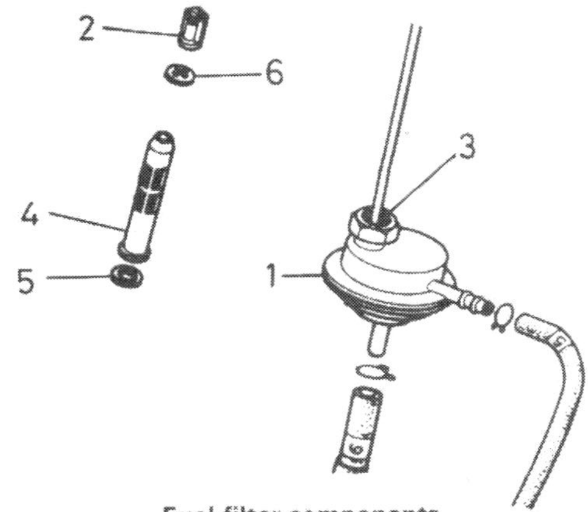

Fuel filter components

1	*Fuel tap*	4	*Filter*
2	*Air filter*	5	*Washer*
3	*Nut*	6	*Circlip*

3 *Air filter cleaning*

The air filter consists of a moulded plastic housing which contains an oil impregnated foam element. Access to this element may be gained after removal of the carburettor shield and the housing cover. The carburettor shield is retained in position by a single cross head screw which passes through its forward edge. The housing cover is retained in position by two cross head screws whose heads are obscured by the outer face of the cover.

Once exposed, the filter element should be carefully eased out of its location in the filter housing. Closely examine the foam of the element for signs of damage, hardening or perishing caused by age. The element must be renewed if any one of these faults is apparent.

If the element is considered to be serviceable, detach the foam from its metal holder and clean it thoroughly in a non-flammable solvent, such as white spirit, whilst gently squeezing it to remove all traces of oil and accumulated dust. After cleaning, squeeze out the foam by pressing between the palms of both hands and then allow a short time for any solvent remaining in the foam to evaporate. Do not wring out the foam as this will cause damage and thus lead to the need for early renewal.

Reimpregnate the foam with clean SAE 10W/30 oil and gently squeeze out any excess oil before refitting the foam in its holder. Refit the filter element into the filter housing and refit the housing cover. Note that great care should be taken when positioning both the element and cover to ensure that no incoming air is allowed to bypass the element or leak in past the cover joint. If this is allowed to happen, it will allow any dirt or dust that is normally retained by the element to find its way into the carburettor and crankcase assemblies; it will also effectively weaken the fuel/air mixture. Refit the carburettor shield.

Note that if the machine is being run in a particularly dusty or moist atmosphere, then it is advisable to increase the frequency of cleaning and reimpregnating the element. Never run the engine without the element fitted. This is because the carburettor is specially jetted to compensate for the addition of this component and the resulting weak mixture will cause overheating of the engine with the probable risk of severe engine damage.

Remove the filter housing retaining screws ...

4 *Choke control filter cleaning*

The choke control filter consists of a moulded plastic base which forms a union for one of the pipes leading from the choke control valve to the carburettor. A rubber cap is fitted over this base and serves both as an air inlet and as a means of housing a small oil-impregnated foam element. Access to this element can be gained by removing the single screw which serves to retain the carburettor shield in position, detaching the shield and then carefully easing the rubber cap free of its plastic base.

Once removed, the filter element should be closely examined for signs of damage, hardening or perishing caused by age.

... to expose the air filter element

The element must be renewed if any one of these faults is apparent. Check also that neither the filter cap nor base are split or in any way damaged.

If the element is considered to be serviceable, then it should be cleaned thoroughly in a non-flammable solvent, such as white spirit, whilst gently squeezing it to remove all traces of oil and accumulated dust. After cleaning, squeeze out the foam by pressing it between two fingers and then allow a short time for any solvent remaining in the foam to evaporate. Do not wring out the foam as this will cause damage and thus lead to the need for early renewal.

Reimpregnate the foam with clean SAE 10W/30 oil and gently squeeze out any excess oil before placing it on a clean piece of paper or rag ready for refitting. Wipe clean the filter base and clean the air inlet passages within the filter cap by blowing them clear with a jet of compressed air. Wipe clean the inside of the filter cap and fit both the cap and element to the filter base. Finally, locate the carburettor shield in position and secure it by fitting and tightening its single retaining screw.

5 Throttle cable adjustment and oil pump synchronization

These two operations must be undertaken jointly because adjustment of either one affects the other. This is because the throttle and oil pump setting are controlled by a single cable from the twistgrip which runs down to a splitter box. From here, two cables emerge, one to the carburettor and the other to the oil pump. To maintain the correct oil feed rate for a given engine speed the pump cable must therefore be reset when the throttle cable is adjusted.

Check the throttle cable for correct adjustment by measuring the amount of free play at the throttle twistgrip flange. If the measurement obtained is outside the set limits of 2 – 6 mm (0.08 – 0.24 in), then the cable must be adjusted by turning the adjuster at the handlebar end of the cable. If the correct amount of free play cannot be obtained by this method of adjustment, then it is recommended that the cable be renewed. Do not omit to retighten the adjuster locknut on completion of adjustment.

Checking the oil pump control cable for correct adjustment will involve gaining access to the pump by removing the centre body panel from the machine. This panel is retained in position by a single dome head nut with washer and a single screw.

The cable can now be checked for adjustment by opening the throttle so far as possible and then checking that the pointer on the pump lever is in exact alignment with the corresponding pointer case in the pump body. If cable adjustment is found to be incorrect, move the cable guide through its retaining bracket until both pointers come into alignment. Do this by rotating the guide retaining nuts in the required direction. With the pointers aligned, secure the guide in its retaining bracket by tightening the two nuts against each other.

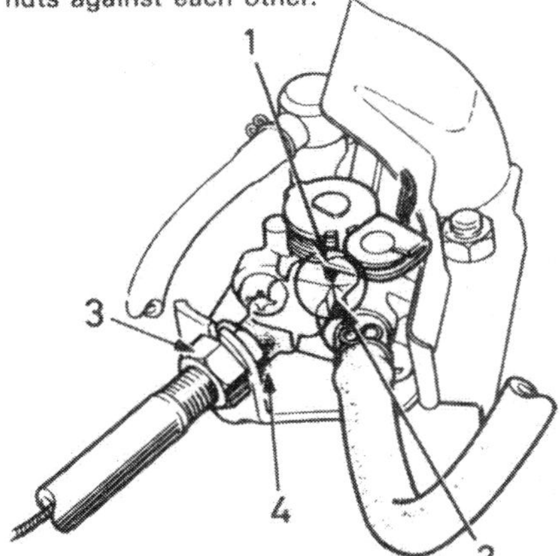

Oil pump cable adjustment synchronisation marks

1 Oil pump alignment mark *3 Adjusting nut*
2 Index mark *4 Locknut*

Expose the choke control filter element

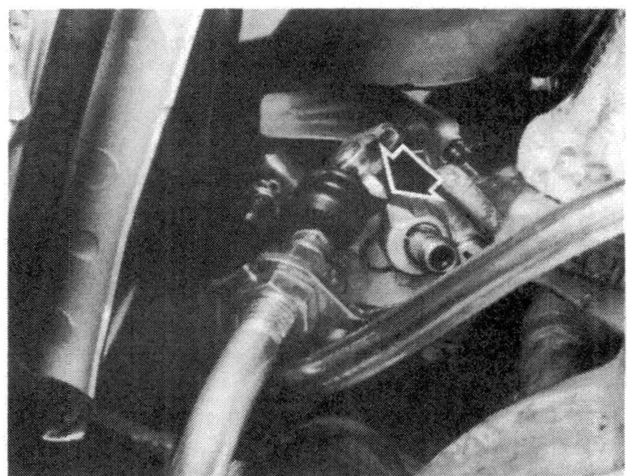

Align the oil pump pointers ...

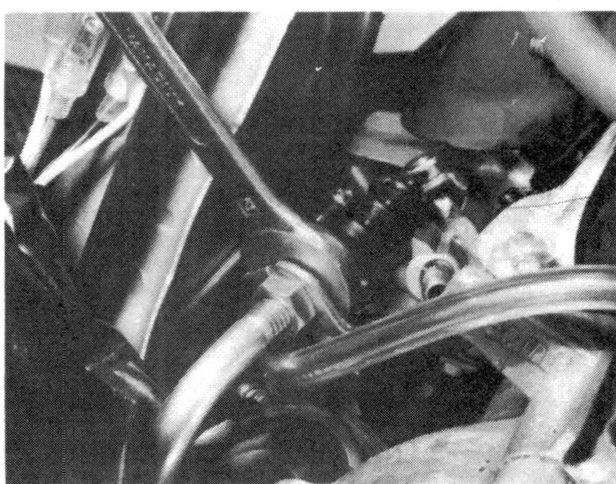

... by rotating the cable guide securing nut

6 Carburettor adjustment

Before checking for correct idle speed adjustment, run the engine until it reaches normal operating temperature, noting that this is best done by riding the machine for a few miles. Identify the pilot air screw and the throttle stop screw. The heads of both of these screws may be seen through the forwardmost slot in the carburettor shield, the pilot air screw being the lower of the two.

Important note: before undertaking any adjustment work with the engine running, it is essential to remember that the Melody has an automatic transmission system. It will be appreciated that as the idle speed rises the centrifugal clutch will engage and the machine will attempt to move off. Be particularly wary of inadvertently 'blipping' the throttle. The machine should be placed squarely on its centre stand, checking that the rear wheel is raised clear of the ground, before work commences.

Start the engine and allow it to idle. Starting from the nominal setting of 1¾ turns out from fully in, turn the pilot air screw clockwise until the engine is heard to falter. Note the position of the screw at this point and then turn it anti-clockwise through the nominal setting position until the engine is once again heard to falter. Again note the position of the screw and then turn it clockwise to a point approximately mid-way between the two noted positions. By this method it is possible to establish the exact setting at which the engine idle speed is fastest and most regular; the setting of the pilot air screw at this point coinciding approximately with its recommended nominal setting. Once the pilot air screw is thus set, turn the throttle stop screw to give a reasonably slow but reliable idle speed.

Always guard against the possibility of incorrect carburettor adjustment which will result in a weak mixture. Two stroke engines are very susceptible to this type of fault which will cause rapid overheating and often subsequent engine seizure.

7 Spark plug examination

Remove the spark plug, using a proper spark plug spanner to avoid any risk of damage to the ceramic insulator. Examine the colour and condition of the electrodes, comparing this with the spark plug condition chart in Chapter 3. This will give an indication of the general condition of the engine. Clean the plug electrodes using a wire brush and a small magneto file or fine emery cloth. If the outer electrode is thin, or the centre electrode has been eroded excessively, the plug must be renewed. The gap can be measured with a feeler gauge, and should be 0.6 to 0.7 mm (0.024 to 0.028 in).

If necessary, adjust the gap by bending the outer electrode. On no account should any attempt be made to bend the inner electrode, or damage to the ceramic insulator nose will almost certainly result. Clean the spark plug threads and wipe them with a trace of graphited grease. Refit the plug by hand, then tighten it carefully with the plug spanner, without over-tightening.

8 Control cable lubrication

The control cable lubrication detail given in the weekly/20 mile service schedule will serve to supplement the full lubrication, which should be carried out as follows.

Disconnect the top of the cable in question, and build up a small funnel of plasticine or similar around the top of the outer cable. Lodge the cable in an upright position, and fill the funnel with light machine oil or engine oil, leave the oil to drain through, preferably overnight.

A quicker and more positive method of lubrication is to use an hydraulic cable oiler which is fairly inexpensive and can be obtained from many motorcycle shops or by mail order from companies advertising in the motorcycle press.

9 Steering and suspension examination

Check the operation of the front and rear suspension, noting any stiffness or free play that might have developed. Whilst this is unlikely, wear will eventually take place and the above symptoms indicate the need for overhaul. The same is true of the steering, which should move freely from lock-to-lock. Any signs of roughness indicate that the bearings are in need of attention, as does any discernible free play. Reference should be made to Chapter 4 for further information on the above items.

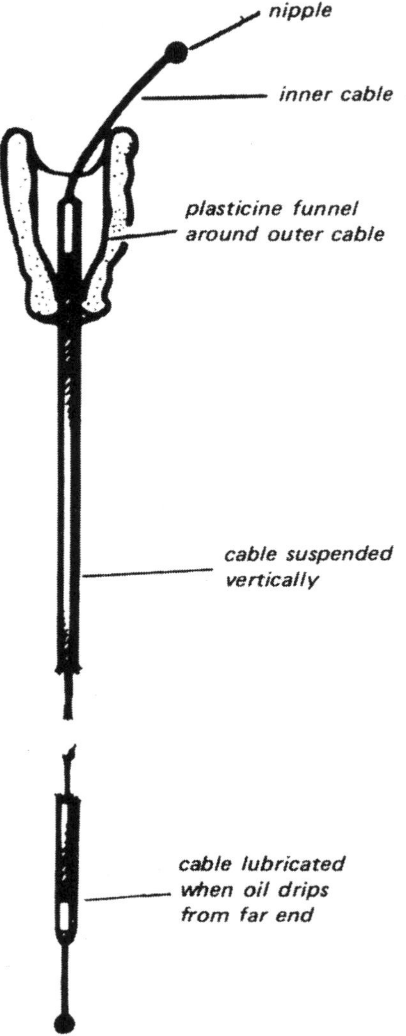

Oiling a control cable

10 Wheel examination

Place the machine on its centre stand and carry out an examination of each wheel as follows. Raise the wheel to be examined clear of the ground by positioning a large wooden block, or similar, either under the footrest panel mounting points or beneath the engine crankcase.

Spin the wheel and check for rim alignment in both the axial and radial planes by placing a pointer close to the rim edge. If the total alignment variation is greater than the given service limit of 2.0 mm (0.08 in), then it is the manufacturer's recommendation that the wheel be renewed. This is, however, a counsel of perfection and in practice a larger amount of runout may not affect the handling qualities of the machine to an excessive degree.

When cleaning the machine, examine the wheel carefully. Any areas where the paint finish has been chipped away must be refinished at the earliest available opportunity if the likelihood of the wheel rusting is to be avoided. Any rust already present on the wheel should be removed by rubbing the affected area with a strip of emery cloth until the exposed metal is bright and clean and then coating the area with one of the rust removing agents sold by any of the well known motor accessory stockists before applying the final paint finish.

Inspect the complete wheel for localised damage in the form of cracks or dents. It is possible that a dent can be ignored as long as the paint finish has remained undamaged, but a crack will render the wheel unfit for further use unless it is found that a permanent repair is possible through welding. This method of repair requires a great degree of skill and therefore the advice of a wheel repair specialist should be sought.

Check closely that the area of wheel rim which supports the tyre bead is in no way damaged. A tyre that is improperly seated and which moves around the wheel rim can only constitute potential danger.

Finally, grasp each side of the wheel and feel for any side-to-side movement. Any movement felt will indicate wear in the wheel bearings (front) or wheel centre spline and/or bearings (rear). Any roughness heard or felt in the bearings as the wheel is rotated will indicate that the bearings are in need of renewal.

11 Brake adjustment and wear check

It is essential that the brakes on any motorcycle are always kept in correct adjustment. The procedure for carrying out adjustment of the brakes on the Honda Melody is both simple and straightforward and identical for each brake. Commence by checking the amount of free play measured at the handlebar lever end. If the amount of play measured is outside the set limit of 10 – 15 mm (0.4 – 0.6 in), then it must be altered by turning the adjuster nut at the brake drum end of the cable. Check for correct operation of the brake by spinning the wheel and applying the brake lever. If the brake shoes are heard to be brushing against the surface of the wheel drum, when the brake is not in use, then back off the adjuster nut slightly until all indication of binding disappears.

Honda state that for the front brake only, it is permissible to compensate for wear on the brake shoe linings by moving the brake cam operating arm in relation to the cam itself. If it is found that it is no longer possible to obtain the correct amount of free play at the handlebar lever end by turning the cable adjuster, then the brake cam operating arm should be detached from the cam shaft and moved one spline in the anti-clockwise direction before being refitted. Note that this operation may only be performed once.

With the brakes correctly adjusted, apply each handlebar lever fully and check the position of the indicator plate mounted on the brake cam shaft with the arrow cast into the brake backplate. If the arrow on the indicator plate is seen to align with or go past the arrow on the brake backplate, then the brake shoe linings have worn beyond their service limits and should be renewed.

12 Reduction gearbox oil level check

Position the machine on an area of flat and level ground and place it on its centre stand. Examine the gearbox casing for any signs of oil leakage. If oil leakage is evident, the cause must be rectified as soon as possible. Allowing the gearbox to operate dry, or with less than the recommended amount of oil, will result in a serious and very expensive failure of its component parts. The dangers of rear wheel seizure whilst the machine is in motion are all too obvious.

Remove the oil level bolt, with its sealing washer, from the cover of the gearbox. If oil starts to issue from the bolt hole, then the oil level is correct. If no trace of oil is apparent, then the oil level is low and the gearbox must be replenished with SAE 10W/30 oil through the oil level hole. Before refitting and tightening the oil level bolt, check that its sealing washer is not split or torn.

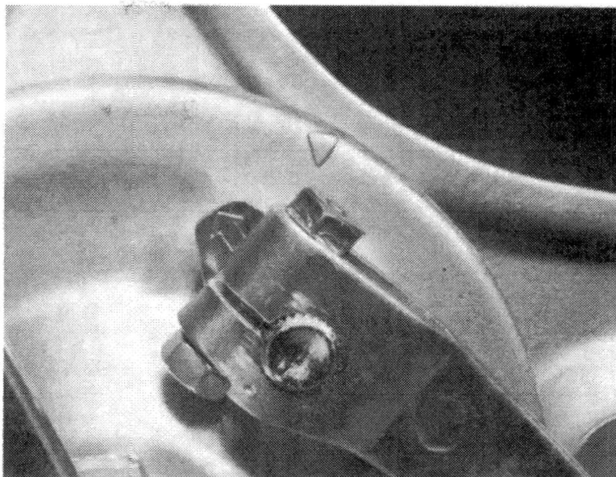

Check the position of the brake indicator plate

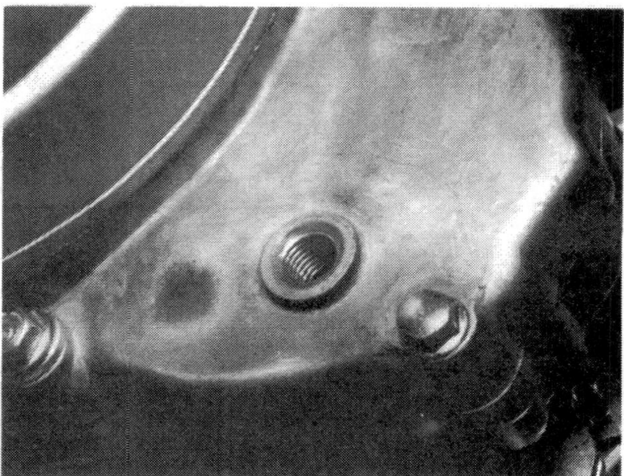

Remove the gearbox oil level bolt

Every two years or 2000 miles (3000 km)

Carry out the service operations listed under the previous headings, then complete the following:

1 Reduction gearbox oil change

Because of the small amount of oil contained in the reduction gearbox and its susceptibility to condensation, it is imperative that the oil be changed on a regular basis. Failure to do this will result in a loss of the lubricating properties of the oil with the resulting high rate of wear between the gearbox component parts.

Before changing the gearbox oil, start the machine and take it for a short ride. This will warm and therefore thin the oil in the gearbox, thereby making it drain more thoroughly. Position the machine on its centre stand on an area of flat and level ground. Place a container of at least 90 cc (3.16 Imp fl oz) capacity beneath the gearbox and remove the gearbox drain bolt together with its sealing washer from the base of the gearbox housing. Whilst the oil is draining, remove the oil level bolt (this is the larger-headed bolt) together with its sealing washer. Inspect and if necessary, renew the two sealing washers.

On completion of the oil draining, refit and tighten the drain bolt with its sealing washer. Refill the gearbox with clean SAE 10W/30 oil by pouring the oil in through the level hole. Approximately 90 cc (3.16 Imp fl oz) of oil should be needed. The oil level is correct when oil is seen to flow back out of the oil level hole. Refit and tighten the oil level bolt with its sealing washer and wipe away any excess oil from the gearbox casing. On completion of the first ride, check around the gearbox for signs of oil leakage. If leakage is evident, the cause must be rectified as soon as possible.

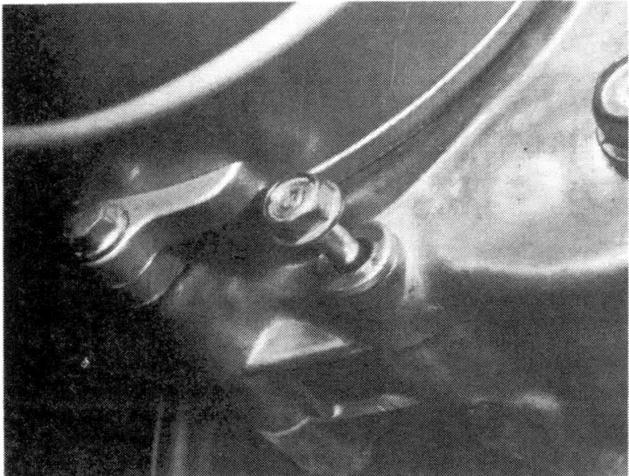
Remove the gearbox drain bolt

2 *Centrifugal clutch shoe examination*

Honda recommend that access be gained to the centrifugal clutch assembly, for the purpose of measuring the amount of wear on the clutch shoe linings, at the above stated mileage interval. This is not really necessary unless it is found that the engine speed is required to be abnormally high before power is transmitted to the rear wheel.

If a defect in the clutch assembly is suspected, then it is advisable to carry out a check of the complete transmission assembly as detailed in Sections 25, 26 and 27 of Chapter 1.

3 *Engine and exhaust system decarbonisation*

For normal decarbonising work, it will only be necessary to remove the cylinder head and exhaust system, together with their associated component parts, whilst leaving the cylinder barrel undisturbed. Full details for doing this are contained in the relevant Chapters of this Manual.

With the piston at the top of the cylinder barrel bore, smear some grease between the edge of the piston crown and the bore surface; this will prevent any carbon deposits from dropping into the bore and make the final cleaning of carbon easier, owing to the fact that the carbon deposits will become stuck to the grease.

Using a hardwood or plastic scraper, remove the carbon deposits from the piston crown. It is desirable to obtain as smooth a finish as is possible, although this is difficult without removing the piston. Take care not to damage the bore surface or piston crown, and avoid the use of sharp instruments, such as screwdrivers, for this reason. The piston crown may, finally, be polished to lessen the likelihood of future carbon build up and to remove any small remaining traces of carbon; metal polish and a buffing wheel or piece of rag may be used for this purpose. Ensure all traces of polish are removed prior to reassembly to prevent premature wear of the piston and bore.

The cylinder head may be dealt with in a similar manner. Any stubborn traces of carbon may be removed by using a fine abrasive paper, although great care must be taken not to remove any great amount of metal from the surface thus altering the shape of the combustion chamber.

The exhaust port must also be cleaned of carbon. A build up of carbon in this area will restrict the flow of exhaust gases from the cylinder.

Before refitting the cylinder head, take care to remove all traces of contamination by washing each component thoroughly with solvent whilst taking care to observe the necessary fire precautions.

Due to the design of the exhaust system and in particular the complex internal substrate of the silencer box, cleaning by conventional means (dissolving deposits with chemicals) is not possible. It is recommended that all carbon deposits be removed from the front end of the exhaust pipe using a suitable scraper and if this fails to improve a blocked system, then the system should be renewed.

In skilled hands, it is possible to split open the exhaust system and burn off any deposits with a welding torch, and then weld up the system. Note, however, that this will have the disadvantage of ruining the paint finish and must only be carried out by someone experienced in this type of work. If the system is severely blocked, it advised that the advice of a Honda dealer is sought.

Additional service items

1 *Cleaning the machine*

Keeping the motorcycle clean should be considered an important part of the routine maintenance, to be carried out whenever the need arises. A machine cleaned regularly will not only succumb less speedily to the inevitable corrosion of external surfaces, and hence maintain its market value, but will be far more approachable when the time comes for maintenance or service work. Furthermore, loose or failing components are more readily spotted when not partially obscured by a mantle of road grime and oil.

The body panels of the Honda Melody are of moulded plastic construction and need to be treated in a different manner from any metal cycle parts when it comes to cleaning. The plastic from which the panels are made will be adversely affected by traditional cleaning and polishing techniques, and lead as a result, to the surface finish deteriorating. Avoid the use of strong detergents which contain bleaching additives, scouring powders or other abrasive cleaning agents, including all but the finest aerosol polishes. Cleaning agents with an abrasive additive will score the surface of the panels thereby making them more receptive to dirt and permanently damaging the surface finish. The most satisfactory method of cleaning the body panels is to 'float' off any dirt from their surface by washing them thoroughly with a mild solution of soapy water and then wiping them dry with a clean chamois leather. A light coat of polish may then be applied to each panel as necessary if it is thought that the panel is beginning to lose its original shine. The remaining areas of the machine can be cleaned by using the procedures described in the following paragraphs.

The plated parts of the machine should require only a wipe with a damp rag. If the plates parts are badly corroded, as may occur during the winter when the roads are salted, it is preferable to use one of the proprietary chrome cleaners. These often have an oily base, which will help to prevent the corrosion from recurring.

If the engine parts are particularly oily, use a cleaning compound such as 'Gunk' or 'Jizer'. Apply the compound whilst the parts are dry and work it in with a brush so that it has the opportunity to penetrate the film of grease and oil. Finish off by washing down liberally with plenty of water, taking care that it does not enter the carburettor or the electrics.

Whenever possible, the machine should be wiped down after it has been used in the wet, so that it is not garaged under damp conditions which will promote rusting. Remember there is little chance of water entering the control cables and causing stiffness of operation if they are lubricated regularly as recommended in the preceding Sections of this Chapter.

2 Oil filter examination and cleaning

Honda recommend that this operation is carried out on a regular routine maintenance basis but do not specify the exact service interval. It is the opinion of the Author that unless the oil contained in the tank has become contaminated, either by excessive condensation or by the introduction of dirt and the like through the filler hole, then there is no need to remove the filter until removal of the oil feed pipe, oil tank or the engine/transmission unit is necessary. The final decision should however be made by the owner. Note that suspected oil starvation problems may well be traced to a blocked filter.

The oil filter comprises a small cylindrical filter element which is contained in a rubber adaptor which itself forms a union between the outlet point of the oil tank and the oil feed pipe to the oil pump. In order to gain access to this element it is first necessary to drain the tank of oil. To do this, obtain a clean container of at least 1.2 litres (2.1 Imp pints) capacity and place it beneath the machine. Remove the centre body panel from the machine by unscrewing its single dome head nut with washer and its single screw. This will give enough access to the oil pump to allow the oil feed pipe to be detached from the pump and its end placed on the container. On completion of draining the oil, unclip the rubber adaptor from the tank outlet and carefully lower the filter element clear of the tank. Note the routing of the pipe and then draw it clear of the machine.

If oil starvation problems have been experienced, yet the filter element is seen to be clear, then the oil feed pipe must be checked for blockage or signs of damage leading to leakage of oil. Renew the pipe, if necessary, and proceed to clean the filter element as follows.

The element should be cleaned by directing a jet of compressed air through its open end as shown in the figure accompanying this text. Do not attempt to direct air through the sides of the element as this will only serve to collapse the fine filter mesh, making renewal of the element necessary. Take care to wear adequate eye protection against any spray back of oil from the element and remember to observe the necessary fire precautions. If necessary, any stubborn traces of contamination can be removed by gentle rubbing of the element with a soft-bristled brush, a used toothbrush is ideal. On completion of cleaning, closely inspect the element for any splits or holes that will allow the passage of sediment through it and into the pump. Renew the element if it is in any way defective.

Fitting of the filter components and oil feed pipe is a direct reversal of the removal procedure, whilst noting the following points. After having refilled the oil tank with Honda 2-stroke injector oil, or equivalent, prime the oil feed pipe and bleed the oil pump of air in accordance with the instructions given in Section 20 of Chapter 2. Do not refit the centre body panel until the machine has been run and a comprehensive check for oil leaks carried out at all disturbed connections.

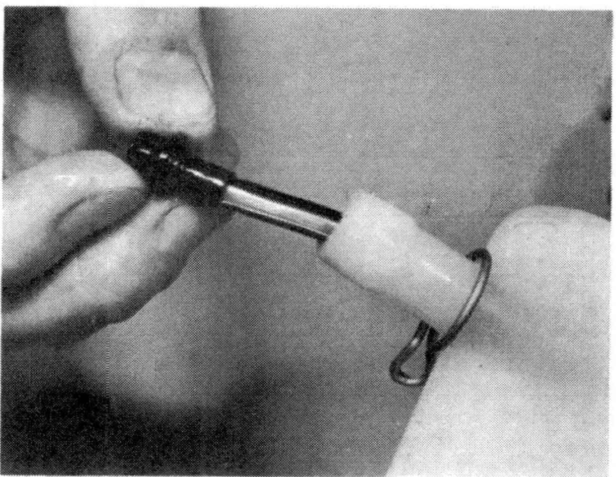

Withdraw the oil filter element

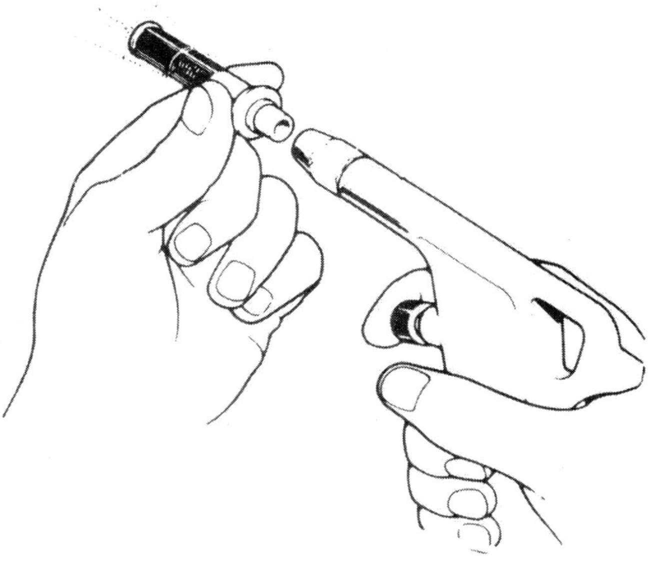

Cleaning the oil filter element

Castrol Engine Oils

Castrol Grand Prix

Castrol Grand Prix 10W/40 four stroke motorcycle oil is a superior quality lubricant designed for air or water cooled four stroke motorcycle engines, operating under all conditions.

Castrol Super TT Two Stroke Oil

Castrol Super TT Two Stroke Oil is a superior quality lubricant specially formulated for high powered Two Stroke engines. It is readily miscible with fuel and contains selective modern additives to provide excellent protection against deposit induced pre-ignition, high temperature ring sticking and scuffing, wear and corrosion.
Castrol Super TT Two Stroke Oil is recommended for use at petrol mixture ratios of up to 50:1.

Castrol R40

Castrol R40 is a castor-based lubricant specially designed for racing and high speed rallying, providing the ultimate in lubrication. Castrol R40 should never be mixed with mineral-based oils, and further additives are unnecessary and undesirable. A specialist oil for limited applications.

Castrol Gear Oils

Castrol Hypoy EP90

An SAE 90 mineral-based extreme pressure multi-purpose gear oil, primarily recommended for the lubrication of conventional hypoid differential units operating under moderate service conditions. Suitable also for some gearbox applications.

Castrol Hypoy Light EP 80W

A mineral-based extreme pressure multi-purpose gear oil with similar applications to Castrol Hypoy but an SAE rating of 80W and suitable where the average ambient temperatures are between 32°F and 10°F. Also recommended for manual transmissions where manufacturers specify an extreme pressure SAE 80 gear oil.

Castrol Hypoy B EP80 and B EP90

Are mineral-based extreme pressure multi-purpose gear oils with similar applications to Castrol Hypoy, operating in average ambient temperatures between 90°F and 32°F. The Castrol Hypoy B range provides added protection for gears operating under very stringent service conditions.

Castrol Greases

Castrol LM Grease

A multi-purpose high melting point lithium-based grease suitable for most automotive applications, including chassis and wheel bearing lubrication.

Castrol MS3 Grease

A high melting point lithium-based grease containing molybdenum disulphide. Suitable for heavy duty chassis application and some CV joints where a lithium-based grease is specified.

Castrol BNS Grease

A bentone-based non melting high temperature grease for ultra severe applications such as race and rally car front wheel bearings.

Other Castrol Products

Castrol Girling Universal Brake and Clutch Fluid

A special high performance brake and clutch fluid with an advanced vapour lock performance. It is the only fluid recommended by Girling Limited and surpasses the performance requirements of the current SAE J1703 Specification and the United States Federal Motor Vehicle Safety Standard No. 116 DOT 3 Specification.
In addition, Castrol Girling Universal Brake and Clutch fluid fully meets the requirements of the major vehicle manufacturers.

Castrol Fork Oil

A specially formulated fluid for the front forks of motorcycles, providing excellent damping and load carrying properties.

Castrol Chain Lubricant

A specially developed motorcycle chain lubricant containing non-drip, anti corrosion and water resistant additives which afford excellent penetration, lubrication and protection of exposed chains.

Castrol Everyman Oil

A light-bodied machine oil containing anti-corrosion additives for both household use and cycle lubrication.

Castrol DWF

A de-watering fluid which displaces moisture, lubricates and protects against corrosion of all metals. Innumerable uses in both car and home. Available in 400gm and 200gm aerosol cans.

Castrol Easing Fluid

A rust releasing fluid for corroded nuts, locks, hinges and all mechanical joints. Also available in 250ml tins.

Castrol Antifreeze

Contains anti-corrosion additives with ethylene glycol. Recommended for the cooling system of all petrol and diesel engines.

Chapter 1 Engine and transmission

Refer to Chapter 7 for information on the NB, ND and NP50 models

Contents

Specifications

Engine

Type	Single cylinder, air cooled two-stroke
Porting	Piston and reed valve
Bore	40 mm (1.57 in)
Stroke	39.3 mm (1.54 in)
Capacity	49 cc (2.99 cu in)
Compression ratio	7.0 : 1
Lubrication	Metered pump system

Piston and rings

Cylinder bore to piston clearance	0.035 – 0.050 mm (0.0013 – 0.0019 in)
Service limit	0.10 mm (0.0039 in)
Piston OD	39.955 – 39.970 mm (1.5730 – 1.5736 in)
Service limit	39.90 mm (1.5709 in)
Gudgeon pin OD	9.994 – 10.0 mm (0.3935 – 0.3937 in)
Service limit	9.970 mm (0.3925 in)
Gudgeon pin to piston clearance	0.002 – 0.012 mm (0.0001 – 0.0005 in)
Service limit	0.040 mm (0.0016 in)
Gudgeon pin hole ID	10.002 – 10.008 mm (0.3938 – 0.3940 in)
Service limit	10.030 mm (0.3949 in)
Ring end gap (fitted)	0.15 – 0.35 mm (0.006 – 0.014 in)
Service limit	0.60 mm (0.024 in)

Cylinder barrel

Standard bore diameter	40.0 – 40.015 mm (1.5748 – 1.5754 in)
Service limit	40.050 mm (1.5768 in)

Crankshaft assembly

Crankshaft runout service limit:	
Left	0.10 mm (0.004 in)
Right	0.15 mm (0.006 in)
Big-end radial clearance service limit	0.05 mm (0.002 in)
Big-end side clearance service limit	0.60 mm (0.024 in)
Small-end eye ID	14.005 – 14.017 mm (0.5514 – 0.5519 in)
Service limit	14.030 mm (0.5524 in)

Kickstart assembly

Kickstart shaft OD	13.957 – 13.984 mm (0.5495 – 0.5506 in)
Service limit	13.90 mm (0.5472 in)
Shaft bush ID	14.016 – 14.051 mm (0.5518 – 0.5532 in)
Service limit	14.10 mm (0.5551 in)
Idle gear pinion shaft OD:	
Left side	15.934 – 15.984 mm (0.6273 – 0.6293 in)
Service limit	15.90 mm (0.630 in)
Right side	11.957 – 11.984 mm (0.4707 – 0.4718 in)
Service limit	11.90 mm (0.4685 in)

Clutch

Type	Automatic centrifugal
Drum ID	107.0 – 107.2 mm (4.21 – 4.22 in)
Service limit	107.5 mm (4.23 in)
Shoe lining thickness	4.0 – 4.1 mm (0.157 – 0.161 in)
Service limit	2.0 mm (0.08 in)
Fixed driven face OD	33.950 – 33.975 mm (1.3366 – 1.3376 in)
Service limit	33.930 mm (1.3358 in)
Sliding driven face ID	34.0 – 34.025 mm (1.3386 – 1.3396 in)
Service limit	34.060 mm (1.3409 in)
Driven face spring free length	93.4 mm (3.68 in)
Service limit	88.0 mm (3.46 in)

Transmission

Type	Automatic variable ratio. Drive from crankshaft pulley to rear wheel via toothed V-belt, dry centrifugal clutch and reduction gearbox
Primary reduction ratio	2.00 – 1.00:1
Final reduction ratio	10.243 : 1
Drive belt width	14.0 mm (0.55 in)
Service limit	12.5 mm (0.492 in)
Drive face boss OD	24.995 – 25.025 mm (0.9841 – 0.9852 in)
Service limit	24.960 mm (0.9827 in)
Sliding drive face bush ID	25.035 – 25.095 mm (0.9856 – 0.9880 in)
Service limit	25.130 mm (0.9894 in)
Weight roller OD	15.92 – 16.08 mm (0.627 – 0.633 in)
Service limit	15.40 mm (0.606 in)
Reduction gearbox oil capacity	90 cc (3.16 Imp fl oz)
Oil type	SAE 10W/30

Torque wrench settings

	lbf ft	kgf m
Cylinder head retaining bolts	7.0 – 9.0	0.9 – 1.2
Flywheel generator rotor retaining nut	25 – 29	3.5 – 4.0
Drive pulley retaining nut	25 – 29	3.5 – 4.0
Clutch drum retaining nut	25 – 29	3.5 – 4.0
Clutch backplate nut	25 – 29	3.5 – 4.0
Clutch cover plate bolts	2.0 – 3.0	0.25 – 0.40

Inlet stub retaining bolts	6.0 – 9.0	0.8 – 1.2
Carburettor retaining nuts	7.0 – 9.0	0.9 – 1.2
Engine mounting bolt retaining nut	25 – 33	3.5 – 4.5
Mounting plate centre nut	14 – 22	2.0 – 3.0
Reduction gearbox drain plug	7.0 – 10	1.0 – 1.4
Rear suspension unit lower bolt	18 – 25	2.5 – 3.5
Rear wheel retaining nut	58 – 72	8.0 – 10.0

1 General description

The engine/transmission unit of the Honda Melody follows the current trend in modern Japanese scooter design in that the unit crankcase is formed by two aluminium alloy castings, the left-hand one of which is extended to act as the transmission casing. The complete unit is pivoted at the front to act as a swinging arm, its movement being controlled by a single suspension unit.

The engine unit itself is a single cylinder two-stroke with reed valve induction and pump fed lubrication, the cylinder barrel being inclined 15 degrees rearwards from the vertical. The crankshaft comprises two shafts and integral full flywheels and is supported by two journal ball main bearings. The flywheels are joined by a press-fit crankpin serving as the journal for the big-end bearing. Both the small-end and big-end bearings are of the caged needle roller type.

The outer face of the right-hand crankcase half houses the flywheel generator assembly. Attached to the rotor of this assembly is a finned nylon cooling fan which forces air through the cowling which fits over the cylinder head and barrel. The left-hand crankcase half houses the transmission components. Attached to the crankshaft is a drive pulley over which runs a toothed V-belt. This belt transmits power to a driven pulley which forms part of a centrifugal clutch. Drive from this clutch is then taken into a reduction gearbox which is housed in the rearmost section of the transmission casing. An intermediate shaft passes power from the clutch shaft to the rear wheel stub-axle. All components housed within the reduction gearbox are lubricated by a separate reservoir of oil contained within the gearbox itself.

Engine starting on all models is by a conventional kick-start system; drive from the segment on the kickstart lever shaft being transmitted via an intermediate reduction pinion to a pinion mounted on the left-hand crankshaft. As well as being fitted with this kickstart arrangement, Honda Melody deluxe models are also equipped with an electric start. The starter motor is mounted on the front of the left-hand crankcase half and drives direct to the crankshaft mounted pinion.

2 Operations with the engine/transmission unit in the frame

The engine/transmision unit is well enclosed when fitted in the frame and this will restrict easy access to some of the ancillary components. The component parts listed below can be removed or worked on without having to resort to the removal of the entire unit, although it may well prove easier to do so. Bodywork panels should be removed when the need dictates.

1 Fan cowlings
2 Cylinder head
3 Cylinder barrel
4 Piston
5 Flywheel generator rotor and stator
6 Air cleaner
7 Carburettor
8 Reed valve
9 Kickstart mechanism
10 Transmission components

If it is found necessary to give attention to either of the crankcase halves or to the crankshaft assembly, big-end or main bearings, then it will be necessary to remove the complete engine/transmission unit from the frame before operation of the crankcase halves is possible. If a number of the operations listed above are to be carried out, then it may be found advantageous to remove the unit from the frame and place it on a sturdy work surface of a convenient height in order to facilitate easy removal and fitting of the components concerned.

3 Removing the engine/transmission unit from the frame

1 Position the machine on its centre stand, preferably on a piece of ground that is both firm and level. Ideally, it is helpful to place the machine on a stout raised platform so that it is raised approximately two feet above ground level. This is not essential but will greatly reduce the amount of stooping that will otherwise be required. Allow plenty of working space around the machine. All of the dismantling work can be carried out unaided. It may, however, be found advantageous to have an assistant present to steady the machine whilst the engine/transmission unit is eased clear of its frame attachment points.

2 Raise the seat and isolate the battery from the electrical system by disconnecting the lead to the battery negative (-) terminal. This simple precaution will ensure that no shorting of exposed contacts occurs whilst disconnecting electrical components during removal of the unit. It is advisable, at this stage, to remove completely the battery from the machine so that it may be properly serviced in accordance with the instructions given in Chapter 6 of this Manual.

3 Remove the one dome head nut with washer and the single screw from the front of the centre body panels and then ease the panel clear of the machine. Locate the single flange nut which retains the rear mudguard to the underside of the rear body panel. This nut is located in the centre of the rear edge of the mudguard. With the nut removed, the mudguard can be eased down and rearwards to free it from its forward locating tabs.

4 Detach the shield from the carburettor and air filter assembly by removing its single retaining screw. Release the fuel pipe retaining clip and pull the pipe off the vacuum-operated fuel tap. With the pipe detached, allow any fuel in it to drain into a small clean container. Take care to observe the necessary fire precautions whilst doing this. Release the vacuum pipe retaining clip and pull the pipe off the vertical stub of the fuel tap.

5 Move to the top of the carburettor and unscrew the top of the mixture chamber. Pull the throttle valve assembly out of the carburettor body and attach it and the cable to a point on the frame which is well clear of the engine unit.

6 Trace the oil feed pipe from the oil tank to the oil pump and release its retaining clip before pulling it clear of the pump. Clamp the end of the pipe to prevent any leakage of oil; a strong bulldog clip is ideal for this purpose. Alternatively, plug the end of the pipe with a bolt of the appropriate shank diameter. Allow the pipe to hang well clear of the engine unit.

7 Before attempting disconnection of the oil pump control cable from the pump lever, it will be necessary to remove the heat guard from the exhaust pipe where it joins the cylinder barrel. This guard is retained in position by one bolt and one nut. With the guard removed, push the oil pump lever towards the cable retaining bracket and disconnect the end nipple from the lever. Release the cable adjuster from its retaining bracket by sliding the rubber gaiter off its threaded end and unscrewing the innermost of the two retaining nuts to allow the adjuster to be pulled outwards and clear of the bracket. Thread the cable clear of the engine unit.

8 Pull the HT lead suppressor cap off the spark plug and attach it to a point on the frame well clear of the engine unit. Move to the left-hand side of the machine and locate the electrical wires which are rotated down the frame structure, just forward of the cylinder head. Unplug the push connectors and the single block connector. The wires leading into these connectors are colour coded; this should be noted for reference when reconnecting. Check that the disconnected wires are free of the frame so that they will not prevent the engine from being moved rearwards. Note that if the engine is to be left removed from the frame for any length of time, then it is well worth masking off the open ends of these connections to prevent any ingress of moisture or dirt.

9 Move to the rear wheel and unscrew the adjuster nut from the threaded end of the rear brake operating cable. Remove the cable anchor plate by unscrewing its simple retaining bolt from the rear of the transmission casing. Move to the front of the transmission casing and detach the cable guide clamp by unscrewing its single retaining bolt. Pull the cable clear of the brake cam operating lever and move it forward to clear the engine. Refit both of the removed bolts, the anchor plate and the adjuster nut in order to prevent lo ss. Retain the lever trunnion.

10 Detach the rear suspension unit from the transmission casing by removing its single retaining bolt and pulling it rearwards to clear the mounting lug. Refit the bolt to the unit to prevent loss and then pull the end of the unit up and back as far as possible before using a length of string to secure it to the luggage rack.

11 Before removing the engine unit pivot bolt and thus freeing the unit from the frame, carry out a final check to ensure that all the disconnected electrical wires, pipes, etc, have been moved well clear of the unit and that there is no possibility of damage being caused through engine components becoming entangled with those fitted to the frame.

12 Unscrew the locknut from the end of the engine unit pivot bolt and remove it together with the plain washer. Using a soft-metal drift and a hammer, carefully drift the pivot bolt from position. With the bolt thus removed, carefully ease the engine unit rearwards until it clears the frame mounting plates. Once free, the unit can be manoeuvred rearwards and to the right so as to clear the machine. Place the engine unit on a prepared work surface, refit the locknut and plain washer to the pivot bolt and store the bolt in a safe place ready for inspection and refitting.

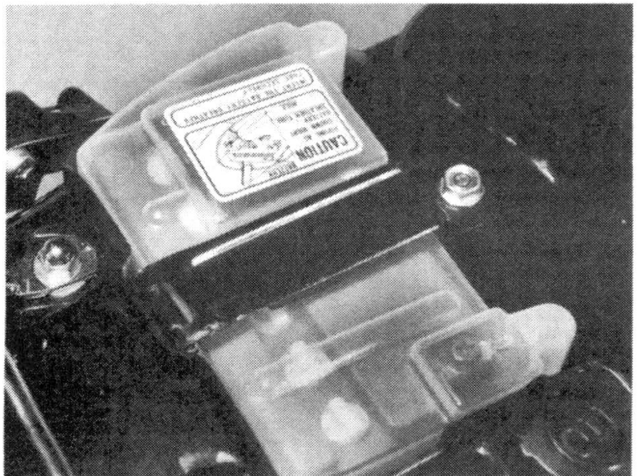
3.2a Remove the battery retaining clamp and cover ...

3.2b ... and disconnect both terminal leads

3.7 Remove the exhaust pipe heat guard

3.8 Unplug the electrical connectors

3.9 Detach the rear brake cable guide

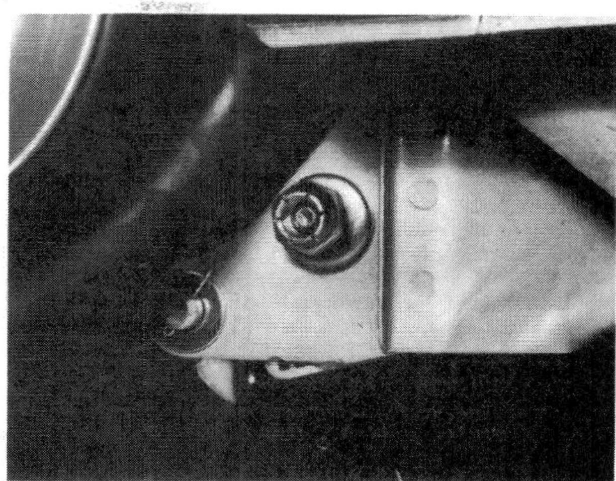

3.12 Remove the engine unit pivot locknut and washer

4 Dismantling the engine/transmission unit: preliminaries

1 Before any dismantling work is undertaken, the external surfaces of the unit should be thoroughly cleaned and degreased. This will prevent the contamination of the engine internals, and will also make working a lot easier and cleaner. A high flash point solvent, such as paraffin (kerosene) can be used, or better still, a proprietary engine degreaser such as Gunk. Use old paintbrushes and toothbrushes to work the solvent into the various recesses of the engine castings. Take care to exclude solvent or water from the electrical components and inlet and exhaust ports. The use of petrol (gasoline) as a cleaning medium should be avoided, because the vapour is explosive and can be toxic if used in a confined space.

2 When clean and dry, arrange the unit on the workbench, leaving a suitable clear area for working. Gather a selection of small containers and plastic bags so that parts can be grouped together in an easily identifiable manner. Some paper and a pen should be on hand to permit notes to be made and labels attached where necessary. A supply of clean rag is also required.

3 Before commencing work, read through the appropriate section so that some idea of the necessary procedure can be gained. When removing the various engine components, it should be noted that great force is seldom required, unless specified. In many cases, a component's reluctance to be removed is indicative of an incorrect approach or removal method. If in any doubt, re-check with the text.

4 It should be noted at this juncture that although no special Honda service tools were required for the purpose of dismantling the engine/transmission unit, it was found necessary to obtain a good quality two-legged puller of the type shown in the photographs accompanying the text of this Chapter in order to remove such items as the flywheel generator rotor, the clutch assembly and the transmission drive pulley. If it is found that this puller is not obtainable through hire or cannot be borrowed, then it is recommended that each component to be removed is inspected closely before deciding on the type of puller to be purchased. Obviously, a puller with a full range of adjustment and one that is equipped with differing types of legs is ideal.

5 Finally, although Honda have greatly reduced the number of crosshead screws used in the construction of their engine units, it was found that certain of these screws used on the Melody required some considerable force to free them. In order to avoid damaging the heads of these screws, it was found necessary to make use of an impact driver.

5 Dismantling the engine/transmission unit: removing the exhaust system

1 Commence removal of the exhaust system by unscrewing the two flange nuts which retain the exhaust pipe to the cylinder barrel. These nuts need only be unscrewed as far as the end of their studs. Remove the heat guard from the silencer by removing the single crosshead screw which serves to retain it in position and then lifting the guard up off its locating clips.

2 Detach the exhaust system from the machine by removing the single bolt, with its plain washer, which passes through the mounting lug on the forward edge of the silencer. Grasp the system at either end and lift it clear of the cylinder barrel studs. Remove and discard the sealing ring which will be left in the recess of the exhaust port. This ring should be replaced with a new item before refitting the exhaust system.

3 Note that full details of cleaning and reprotecting the exhaust system are contained in Section 15 of Chapter 2. Now is the ideal time for carrying out these tasks.

5.1 Remove the exhaust silencer heat guard

6 Dismantling the engine/transmission unit: removing the rear wheel

1 Removing the rear wheel with the engine out of the frame is most easily accomplished with the aid of an assistant. The problem which will be encountered is rotation of the wheel directly any turning force is imposed on the nut in an attempt to loosen it. This problem is easily overcome if the assistant grips the tyre to prevent the wheel from rotating whilst the nut is removed. With the nut thus removed, displace the thick plate washer from the threaded end of the stub-axle and lift the rear wheel off the axle splines. Refit the nut and washer to the axle, both to prevent loss and to protect the end of the axle from damage. Note the presence of the thrust collar contained within the recess of the transmission housing.
2 Note that full details of examining and renovating the rear brake assembly are contained within Chapter 5 of this Manual. It is advisable to do any necessary work on the brake assembly during the general examination and renovation carried out on the crankcase castings.

7 Dismantling the engine/transmission unit: removing the cylinder head cowling and fan assembly

1 The cylinder head and barrel of the engine unit are cooled by means of air being forced between the head and barrel finning and a metal cowling which is bolted in position over both the head and barrel. Air is drawn in through a plastic grid by a fan which is attached to the flywheel generator rotor. It is then forced up through a plastic trunking into the cowling assembly whereby it acts to cool the cylinder head and barrel before being released through vents in the cowling. This means of forced air cooling is essential when the cylinder head and barrel of the engine are almost totally obscured by body panelling, as is the case with the Honda Melody. It must be realised that any attempt to run the machine with part of the cowling removed will almost certainly result in overheating of the engine with the subsequent risk of seizure.
2 Removal of the cowling assembly is largely a matter of working in a logical sequence, commencing with the two cross head screws which retain the plastic grid in position and then detaching the trunking which should be followed in turn by each of the two cowling halves. Note that it will be necessary to detach the choke control valve from the cylinder head to facilitate removal of the left-hand half of the cowling. Commence removal of this valve by disconnecting each of its three pipes from their retaining stubs on the carburettor body and intake adaptor. Do this by releasing each pipe retaining clip before pulling the pipe from position. It will be seen that each one of the three pipe retaining stubs on the valve body is numbered. These numbers correspond with identical numbers cast in the bodies of both the carburettor and intake adaptor. As long as each pipe is left connected to either one of its retaining stubs, there should be no confusion as to its fitted position when the time comes for reconnection of the valve. Detach the valve from the cylinder head by removing each of its two retaining bolts.
3 The nylon cooling fan is secured to the flywheel generator rotor by four crosshead screws. In practice, it was found that these screws were extremely tight and required the use of an impact driver to free them.

8 Dismantling the engine/transmission unit: removing the carburettor and reed valve assembly

1 Commence removal of the carburettor by unscrewing the single bolt which serves to retain the air filter housing in position. This bolt is located beneath the housing, just to the rear of the carburettor float chamber. With the bolt removed, pull the filter housing back to clear the carburettor mouth before lifting it clear of the engine unit.
2 The carburettor can now be freed from the intake adaptor by removing its two retaining bolts. The intake adaptor can be removed by first disconnecting the oil feed pipe from the forward facing stub on the adaptor and then removing each of the four retaining bolts. The reed valve assembly will be revealed directly the intake adaptor is lifted from position. Ease the valve from position and store it in a clean, dry storage space until required for inspection and reassembly.

6.1 Detach the rear wheel

7.2a Remove the cooling grid ...

7.2b ... followed by the trunking ...

7.2c ... and each of the two cowling halves

7.3 The cooling fan is attached to the flywheel rotor

8.1 A single bolt (arrow) retains the air filter housing in position

9 Dismantling the engine/transmission unit: removing the crankshaft pulley and centrifugal clutch

1 In order to expose the crankshaft-mounted drive pulley, the centrifugal clutch and the drive belt, it is necessary to remove the kickstart lever and then the transmission casing cover from the left-hand side of the engine unit.

2 The kickstart lever is retained on its splined shaft by means of a single pinch bolt. With this bolt removed, the lever may be pulled off its shaft. Refit the bolt in the lever to avoid loss.

3 The transmission casing cover is retained in position by nine bolts of various lengths. To avoid the risk of these bolts being refitted in the wrong locations, form a template of the cover from a piece of card and insert each bolt into the card as it is removed. Loosen these bolts evenly and in a diagonal sequence; this will greatly reduce the risk of the casing becoming distorted. With all the bolts removed, tap around the edge of the casing with a soft-faced hammer to break the gasket seal and then pull the casing clear of its two locating dowels. Remove each of these dowels and place them in a safe storage space, ready for reassembly. Take care not to lose the detachable bush contained within the casing.

4 Commence removal of the crankshaft-mounted drive pulley by placing a spanner or socket on both the pulley retaining nut and the flywheel generator rotor retaining nut. Turn one nut against the other. It is quite possible that the rotor retaining nut will loosen before the pulley nut, in which case the rotor will have to be prevented from turning by the following method.

5 Fit a two-legged puller to the generator rotor, as for rotor removal. Make sure that the rotor retaining nut is positioned so as to protect the threaded end of the crankshaft and that the feet of the puller legs are a good fit in the slots of the rotor. Pass a length of hardwood between one of the puller legs and the central column of the puller. Do not use anything made of metal as this will damage the threads of the column. Position the length of wood as near to the rotor as possible and use it as a lever with which to prevent the rotor from turning whilst the drive pulley retaining nut is undone. During this operation exercise great care to prevent the pulley from slipping and damaging the rotor.

6 With the drive pulley retaining nut thus removed, the outer face of the pulley may be pulled clear of the crankshaft end and the drive belt manouevred clear of the crankshaft and centrifugal clutch. Do not impose any more sideways strain on the belt than is absolutely necessary.

7 Strip the gasket away from the transmission casing and discard it. Note the thrust washer which should have remained on the kickstart lever shaft. This washer should now be removed and stored ready for reassembly.

8 Grasp the remaining half of the drive pulley assembly and pull it off the crankshaft end. In order to loosen the nut which retains the drum of the centrifugal clutch in position, it is first necessary to devise some means of preventing the drum from rotating directly a turning force is imposed on the nut. In practice, a method similar to that used for preventing the flywheel generator rotor from turning was used (see paragraph 5 of this Section). Bear in mind that a ring spanner must be fitted over the nut before the puller is fitted in position. Once its retaining nut is removed, the clutch drum may be pulled off its splined shaft. The remaining part of the clutch assembly can now be pulled off its shaft and placed in a clean storage space ready for examination and, if necessary, renovation. Take great care not to allow the shoe linings of this assembly to become contaminated with oil or grease. Finally, it is a good idea to temporarily refit the clutch retaining nut. This will prevent the threaded end of the shaft from becoming damaged during subsequent dismantling of the engine/transmission unit.

9.8 Loosen the clutch drum retaining nut

10 Dismantling the engine transmission unit: removing the kickstart components and starter motor

Melody deluxe

1 Commence removal of the kickstart components by drawing the starter pinion off the left-hand crankshaft. Protect the threaded end of the crankshaft by refitting the drive pulley retaining nut and then place a two-legged puller in position over the crankshaft end. Ensure that the legs of the puller are correctly engaged with the holes cut in the pinion before finally tightening down on the puller thread. In practice, it was found that the pinion was extremely tight on the crankshaft and required some considerable effort to release it. The method used was to screw in on the puller as much as possible without risking any damage to its threaded components and then to give the end of the puller a sharp tap with a heavy hammer before tightening it a little further. Upon repeating this process for the third time, the pinion eventually separated from the crankshaft. Removal of the starter pinion using this method requires extreme care if damage to the pinion or crankshaft end is to be avoided. If the pinion proves stubborn and there is doubt about the successful outcome of this operation it is strongly recommended that the partially dismantled engine be returned to a Honda service agent who will be able to remove the pinion without risk of damage. In order to allow full withdrawal of the pinion from the crankshaft, it is necessary to pull the outer end of the idler pinion shaft away from the pinion. This will prevent the teeth of the pinion from snagging on the edge of the shaft cutout.

Melody (non-starter motor model)

2 This model is fitted on the crankshaft end with a kickstart driven gear only. The procedure for removal of this pinion and the remarks concerning caution in the use of a puller are suitable to those given in the preceding paragraph. The legs of the puller must be fitted outside the pinion so that the feet locate securely on the pinion rear face.

All models

3 Using a pair of pliers grasp the leg of the kickstart return spring and carefully ease its hooked end clear of its retaining plate. Some effort will be required to do this as the spring is very strong. Do not risk placing a finger between the spring leg and the plate whilst detaching the leg as one slip could well result in, at the very least, a badly bruised finger. The kickstart lever shaft, together with the spring, can now be eased out of position. Take note of the pin contained in the section of shaft adjacent to the toothed quadrant. It is possible that directly the end of the return spring is detached from this pin, the pin will drop out of its location and be lost. Prevent this from happening by drawing the pin out of the shaft and placing it in safe storage. Do the same with the bush which will be left in the transmission casing.

4 The idler pinion, together with its shaft, may now be pulled from position. Ensure that the end of the friction spring does not become jammed in the guide channel of the transmission casing and take care to retain the small locating pin which is located in the section of shaft beneath the gear pinion.

5 The kickstart return stop is attached to the transmission casing by a single bolt which holds its retaining plate in position. Remove this bolt, lift the plate from position and pull the stop upwards off its locating stub. Place the complete assembly in safe storage, ready for inspection.

6 On models equipped with a starter motor, the motor should now be detached from the transmission casing by removing the two retaining bolts and then pulling it from position. It may be found that the motor has become quite firmly stuck in the casing, in which case it should be gently tapped from position with a soft-faced hammer. Do not attempt to lever the motor away from the casing as this will only cause damage to the mating surfaces.

7 Finally, it is a good idea to temporarily refit the crankshaft pulley retaining nut to the crankshaft. This will prevent the threaded end of the shaft from becoming damaged during subsequent dismantling of the engine/transmission unit.

10.1 Pull the starter pinion off the crankshaft end

10.2 The kickstart lever shaft bush will remain in the transmission casing

10.5 Pull the starter motor from its housing

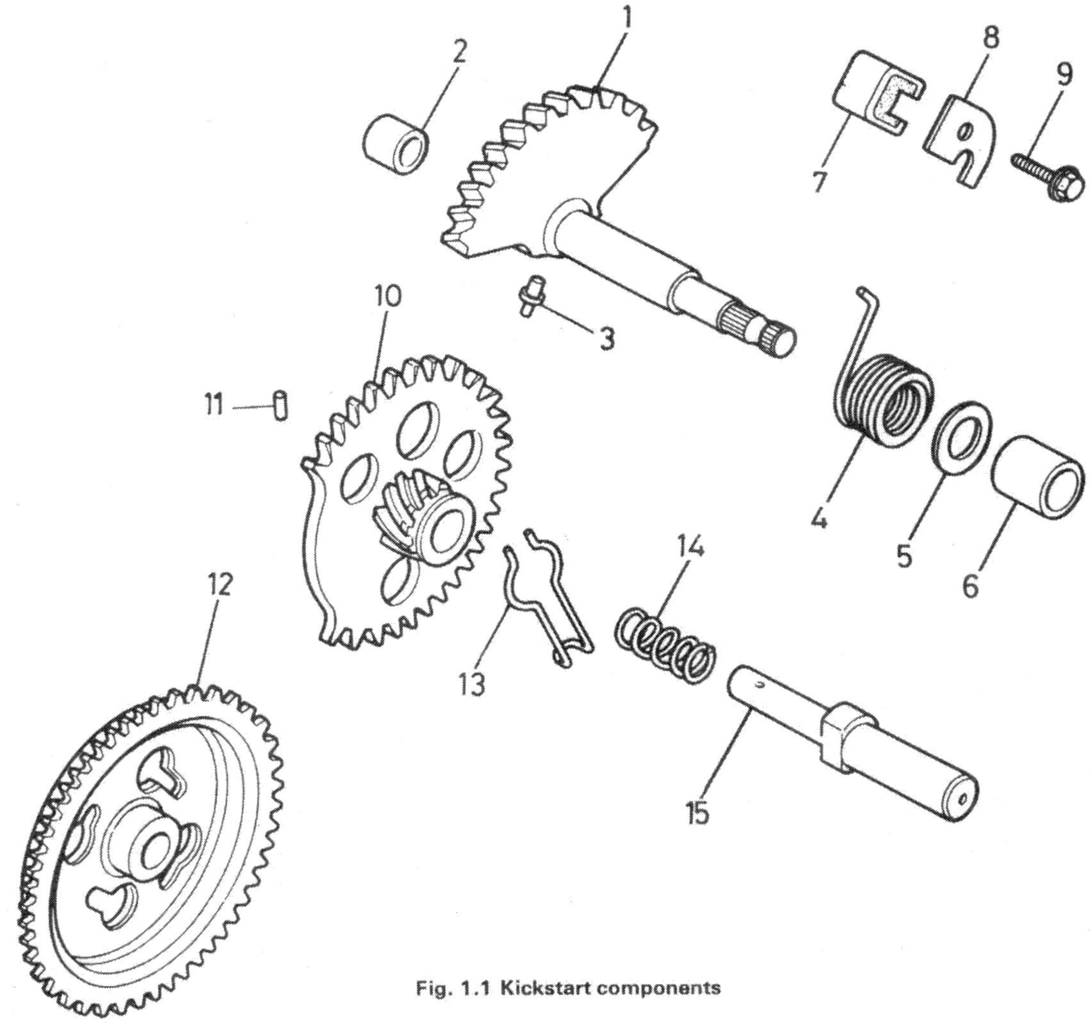

Fig. 1.1 Kickstart components

1 Kickstart lever shaft
2 Bush
3 Pin
4 Return spring
5 Thrust washer
6 Bush
7 Return spring stop
8 Retaining plate
9 Bolt
10 Idler pinion
11 Locating pin
12 Starter pinion
13 Friction spring
14 Spring
15 Idler pinion shaft

11 Dismantling the engine/transmission unit: removing the flywheel generator

1 The flywheel generator will have been exposed by the removal of the air intake grid and trunking as described in Section 7 of this Chapter. Removal of the nylon cooling fan is also necessary to gain access to the rotor retaining nut. It is possible that this nut may have been loosened during the procedure described in paragraph 4 of Section 9 of this Chapter. If this was not the case then it will be necessary to prevent the rotor from turning whilst the nut is loosened by employing the following procedure.

2 Place a close-fitting ring spanner over the rotor retaining nut. Fit a two-legged puller to the rotor, making sure that the feet of the puller legs are a good fit in the slots of the rotor before finally locking the puller in position. Pass a length of hardwood between one of the puller legs and the central column of the pillar. Do not use anything made of metal as this will damage the threads of the column. Position the length of wood as near to the rotor as possible and use it as a lever with which to prevent the rotor from turning whilst the retaining nut is undone.

3 Once the rotor retaining nut has been loosened, move it to the end of crankshaft thread so that its upper face is flush with the end of the crankshaft. This is to prevent any chance of damage occurring to the thread during rotor removal. Withdraw the length of wood and position the ring spanner so that its end is clear of the retaining nut.

4 In practice, it was found that the rotor was very tight on its crankshaft taper and required some considerable force to release it. The method used was to screw in on the puller as much as possible without risking any damage to its threaded components and then to give the end of the puller a sharp tap with a heavy hammer before tightening it a little further. Repeating this process several times eventually led to the rotor becoming free from the crankshaft. The cautionary remarks relating to starter pinion removal which were made in the previous Section, relate equally to this operation,

5 Complete the removal of the rotor by detaching the puller, removing the retaining nut and then lifting the rotor from position to reveal the stator assembly located beneath it. Use the flat of a small screwdriver to ease the Woodruff key out of its keyway in the crankshaft taper and place the key in safe storage until required for examination and refitting.

6 Removal of the stator assembly is only necessary if it is intended to recondition the unit, or if separation of the crankcase halves is considered necessary; the heads of four of the crankcase securing bolts are partially or completely obscured by the stator plate. Note that the presence of oil in the stator housing will almost certainly point to failure of the crankshaft oil seal. If removal of the stator is necessary, unscrew its two retaining bolts and lift it clear of its mounting points. Push out of its location in the stator housing the grommet through which the stator wires pass, and carefully thread the wires clear of the engine unit. Note that the stator plate is not adjustable in relation to the housing and so need not be marked. Once removed, the stator assembly must be wrapped in a clean piece of rag or paper to protect it from contamination and then placed in a clean, dry storage space; the same applies to the rotor. Finally, it is a good idea to refit temporarily the rotor retaining nut. This will prevent the threaded end of the crankshaft from becoming damaged during subsequent dismantling of the engine/transmission unit.

12 Dismantling the engine/transmission unit: removing the reduction gearbox components

1 Before removal of the components contained within the reduction gearbox can take place, the unit must be drained of oil. To do this, place the engine/transmission unit on its side so that the oil drain bolt is facing upwards. Remove this bolt, together with its sealing washer, from its location at the base of the gearbox housing. Do not confuse the drain bolt with the larger oil level bolt. Invert the unit and allow the oil to drain into a container of at least 90 cc (3.16 Imp fl oz) capacity.

2 The gearbox cover can now be detached from the left-hand side of the unit by removing its four retaining bolts and then gently tapping the end of the rear wheel stub-axle with a soft-faced hammer so that the complete gearbox assembly moves away from the gearbox casing. Take the precaution of leaving the rear wheel retaining nut fitted over the threaded end of the stub-axle in order to protect it from damage. Directly the stub-axle becomes free from its bearing in the casing, remove the nut

11.4 Use a puller to remove the flywheel rotor

11.6 Four crankcase securing bolts are located in the stator housing

and carefully ease the gearbox assembly away from the casing whilst making sure that none of the component parts are allowed to slip from position. Place the assembly in a clean, safe storage space. Prevent the possible ingress of dirt into the moving components by covering them with a piece of clean rag or paper.

3 Finally, peel the sealing gasket off the gearbox casing to cover mating surface and discard it. Place the four cover retaining bolts, together with the oil drain bolt, in safe storage until required for reassembly. The same applies to the two gearbox cover locating dowels.

13 Dismantling the engine/transmission unit: removing the unit mounting plate

1 The engine/transmission unit mounting plate serves to link the engine crankcase to the point on the frame about which the unit pivots. It is of pressed-steel construction, its forward end containing two bonded-rubber bushes. The plate is attached to the crankcase by a bolt which passes through the rear of the plate and throught bonded-rubber bushes which are pressed into lugs cast into the crankcase. The plate is prevented from pivoting around this rear attachment by a centre mounting which takes the form of a plate which is slotted into the underside of the crankcase. Passing through this centre mounting is a stud which also passes through the unit mounting plate. The end of the stud is threaded to accept a retaining nut. Finally, in order to prevent metal-to-metal contact between the mounting plate and the crankcase, a rubber buffer is fitted between the two components.

2 Before inverting the engine unit to facilitate removal of the mounting plate, unscrew and remove the spark plug from the cylinder head. Failure to do this will almost certainly result in the porcelain insulator of the plug becoming cracked directly the weight of the engine unit is allowed to bear upon it.

3 Removal of the mounting plate from the crankcase is a straightforward procedure. Commence by removing the nut from the centre mounting stud, together with the plate washer and rubber buffer fitted beneath it. Remove both the nut and washer from the end of the rear mounting bolt and then carefully drift the bolt from position by using a soft-metal drift and hammer. The mounting plate can now be lifted clear of the crankcase and placed to one side, ready for examination. Pull the rubber buffer off the centre mounting stud and detach the centre mounting plate from its retaining slots in the crankcase.

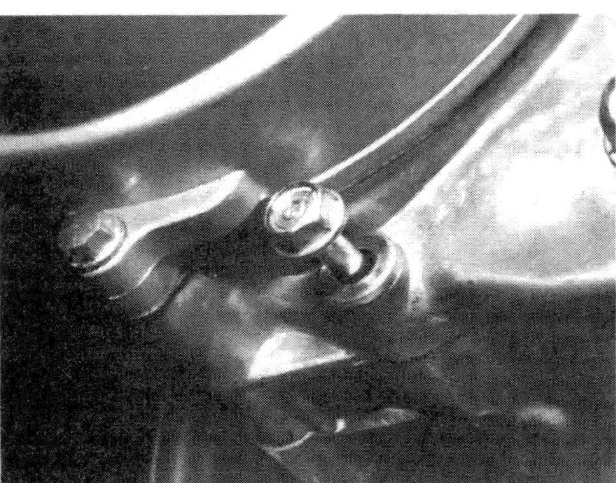
12.1 Remove the gearbox oil drain bolt

13.3a Remove the mounting plate centre nut

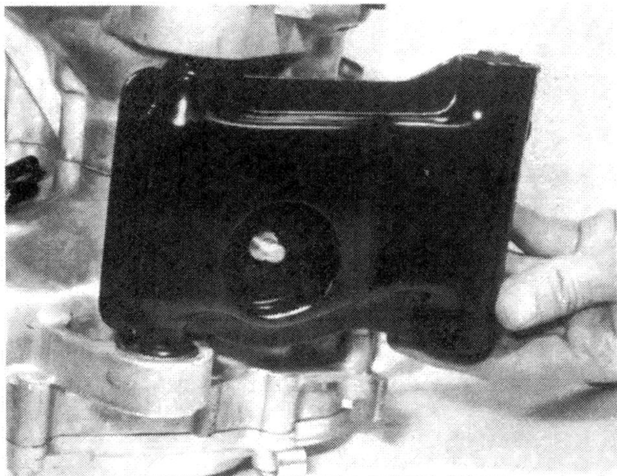
13.3b Detach the mounting plate ...

13.3c ... to expose the rubber buffer and centre mounting stud

14 Dismantling the engine/transmission unit: removing the oil pump

1 The oil pump is secured to the front of the right-hand crankcase half by a single retaining bolt. With this bolt removed, the pump can be pulled clear of its location in the crankcase. Full details of examination and renovation of the pump are contained in Section 19 of Chapter 2. Before wrapping the pump in a clean piece of rag or paper and placing it aside, ready for reassembly, note the condition of the O-ring fitted to the pump boss. This O-ring must be replaced with a new item if it is seen to be in any way damaged or has become flattened during use.

14.1 The oil pump is retained by a single bolt

15 Dismantling the engine/transmission unit: removing the cylinder head, barrel and piston

1 Slacken, evenly and in a diagonal sequence, the four bolts that secure the cylinder head and barrel in position. Remove the bolts and lift off the cylinder head. Discard the cylinder head gasket, a new gasket must be fitted on reassembly to ensure a gas tight seal. In the event of the cylinder head being stuck in position, it should be tapped smartly around its base with a soft-faced hammer to break the seal. Avoid causing damage to the cooling fins whilst doing this.

2 It was found in practice that the cylinder barrel had become firmly bonded to the crankcase and needed some effort to free it. The method used was to place the end of a short length of soft wood against the base of the barrel before tapping the wood with a heavy hammer to lift the barrel away from the crankcase. Take care, whilst doing this, to place the end of the wood against a strengthened section of the barrel, such as the exhaust port. Move around the base of the barrel to ensure that the complete seal is broken and take care not to allow the end of the wood to slip as it is struck. Do not, under any circumstances, attempt to lever the barrel clear of the crankcase.

3 Once freed, lift the cyinder barrel carefully away from the crankcase whilst taking care to catch the piston as it emerges from the bore, thus preventing it from coming into sharp contact with the mouth of the crankcase. As a precaution against pieces of broken piston ring entering the crankcase, a piece of clean rag should be used to pack the open area of crankcase mouth before the piston rings are allowed to leave the confines of the cylinder bore.

4 The piston is supported on a gudgeon pin which is held in place by a wire circlip at each end of the piston bore. The circlips can be removed by grasping the projecting tang with the pointed-nose pliers and pulling the clip from its groove. Alternatively, use the flat of a small screwdriver to dislodge the circlips, as shown in the photograph accompanying this text. This will strain the clips and for this reason they should be renewed as a precaution against their subsequent failure. Once the clips have been removed the piston can be freed by pushing the gudgeon pin from its bore. The pin is usually a light sliding fit and easily removed, but if it appears stuck it can be released as follows.

5 Obtain a container of boiling water in which some rag has been placed. Taking care to avoid scalding, wrap the rag around the piston and leave it for about a minute. The alloy piston will expand much faster than the steel gudeon pin, which will now push out quite easily.

6 Should it ever prove necessary, a drawbolt extractor can be used to remove the gudgeon pin. See the accompanying figure. Never attempt to drift the pin out because there is a real risk of danger to the connecting rod or bearings. As the piston is removed retrieve the caged needle small-end bearing. Note the 'Ex' mark on the piston crown. This must face the exhaust part when the piston is refitted. Finally, remove and discard the cylinder barrel base gasket.

15.4a Remove the gudgeon pin retaining clips ...

15.4b ... to allow withdrawal of the pin

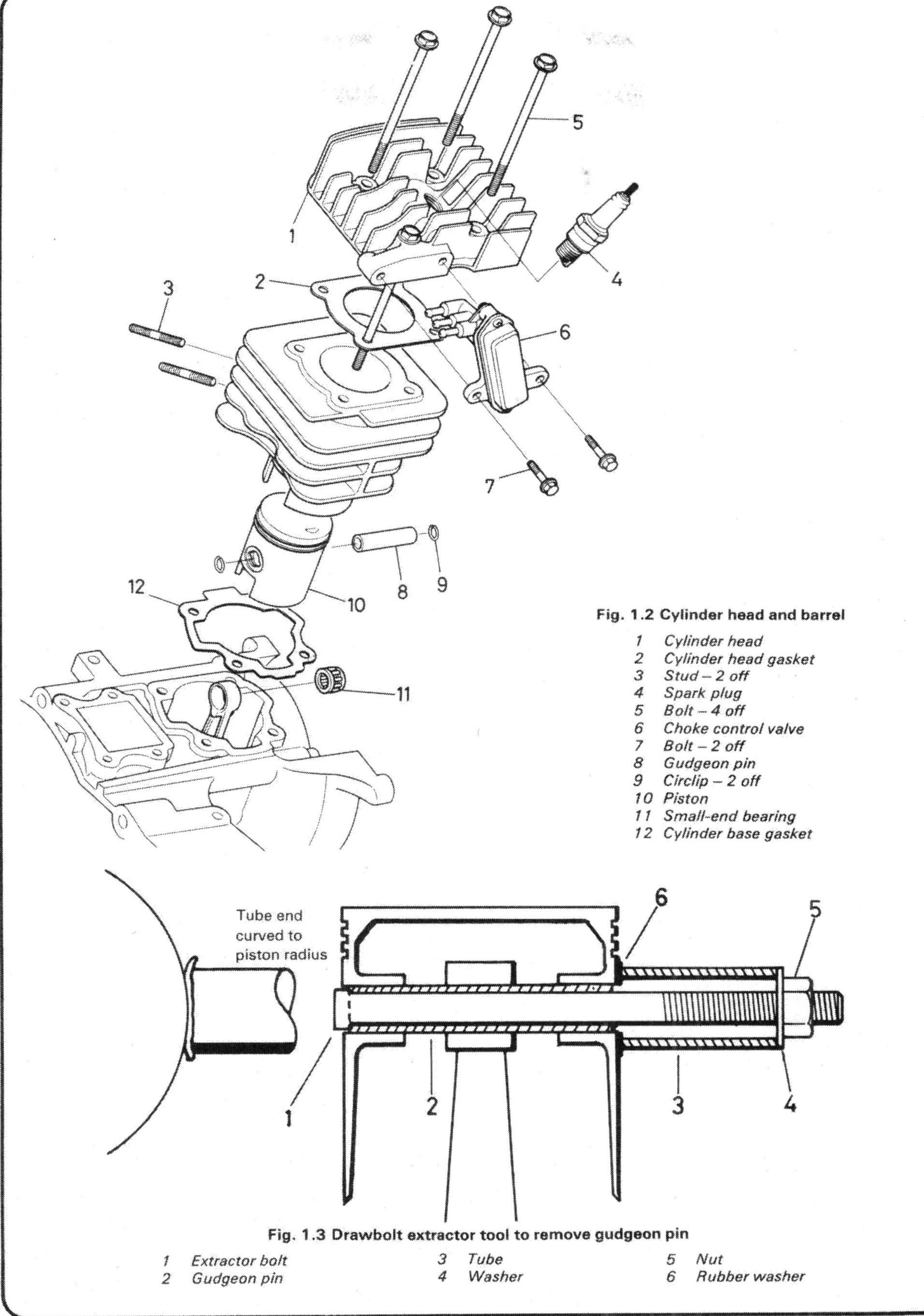

Fig. 1.2 Cylinder head and barrel

1 *Cylinder head*
2 *Cylinder head gasket*
3 *Stud – 2 off*
4 *Spark plug*
5 *Bolt – 4 off*
6 *Choke control valve*
7 *Bolt – 2 off*
8 *Gudgeon pin*
9 *Circlip – 2 off*
10 *Piston*
11 *Small-end bearing*
12 *Cylinder base gasket*

Fig. 1.3 Drawbolt extractor tool to remove gudgeon pin

1 *Extractor bolt*
2 *Gudgeon pin*
3 *Tube*
4 *Washer*
5 *Nut*
6 *Rubber washer*

16 Dismantling the engine/transmission unit: separating the crankcase halves

1 Commence separation of the crankcase halves by positioning the engine unit on a clean work surface so that its right-hand side is uppermost. Check before doing this, that each end of the crankshaft remains protected by the flywheel generator rotor retaining nut or the drive pulley retaining nut. Remove the six crankcase securing bolts. These bolts should be loosened a little at a time at first and in a diagonal sequence; this will preclude any risk of the crankcase castings becoming distorted. Note that two of these bolts are longer than the other four. Take careful note of their fitted position for reference when refitting.
2 Reposition the engine unit so that the crankcase mouth is facing uppermost. Grip the transmission casing with one hand and firmly strike the left-hand end of the crankshaft with a soft-faced hammer. It was found that this blow was enough to cause initial separation of the crankcase halves. Subsequent blows caused further separation, resulting finally in the crankshaft, with the right-hand crankcase half, becoming detached from the transmission casing. This proved to be a fairly easy task which required no excessive force. It is most important to realise the danger of striking the crankshaft end with excessive force as this will almost certainly cause the crankshaft to become distorted.
3 The right-hand crankcase half should now be well supported on wooden blocks so that the crankshaft is in a vertical plane and its left-hand end is raised clear of the surface of the bench. Strike the right-hand end of the crankshaft so as to drive it downwards and free of the crankcase half.
4 Remove the two locating dowels from the crankcase mating surface and place them in safe storage until required for reassembly. Remove the gasket from the mating surface and discard it. The crankshaft assembly should be wrapped in a piece of clean rag or paper, to prevent any ingress of dirt into the big-end bearing, before being placed to one side ready for examination and, if necessary, renovation.
5 It should be noted that Honda recommend the use of a special service tool (No 07935-GA70000), for separating the crankcase halves. If difficulty is experienced when carrying out the procedure described in the above paragraphs of this Section, then it is advisable either to attempt to hire this tool from a local Honda service agent or to return the complete crankcase unit to the agent so that it may be pulled apart and the crankshaft and associated bearings inspected and renovated for a reasonable charge.

17 Examination and renovation: general

1 Before examining the parts of the dismantled engine unit for wear it is essential that they should be cleaned thoroughly. Use a petrol/paraffin mix or a high flash-point solvent to remove all traces of old oil and sludge which may have accumulated within the engine. Where petrol is included in the cleaning agent normal fire precautions should be taken and cleaning should be carried out in a well ventilated place.
2 Examine the crankcase castings for cracks or other signs of damage. If a crack is discovered it will require a specialist repair.
3 Examine carefully each part to determine the extent of wear, checking with the tolerance figures listed in the Specifications sections of this Chapter or in the main text. If there is any doubt about the condition of a particular component, play safe and renew.
4 Use a clean lint free rag for cleaning and drying the various components. This will obviate the risk of small particles obstructing the internal oilways, and causing the lubrication system to fail.
5 Should any studs or internal threads require repair, now is the appropriate time to attend to them. Where internal threads are stripped or badly worn, it is preferable to use a thread insert, rather than tap oversize. Most dealers can provide a thread reclaiming service by the use of Helicoil thread inserts. They enable the original component to be re-used.
6 Sheared studs or screws can usually be removed wth screw extractors, which consist of tapered, left-hand thread screws, of very hard steel. These are inserted by screwing anticlockwise, into a pre-drilled hole in the stud, and usually succeed in dislodging the most stubborn stud or screw. The only alternative to this is spark erosion, but as this is a very limited, specialised facility, it will probably be unavailable to most owners. It is wise, however, to consult a professional engineering firm before condemning an otherwise sound casing. Many of these firms advertise regularly in the motorcycle papers.

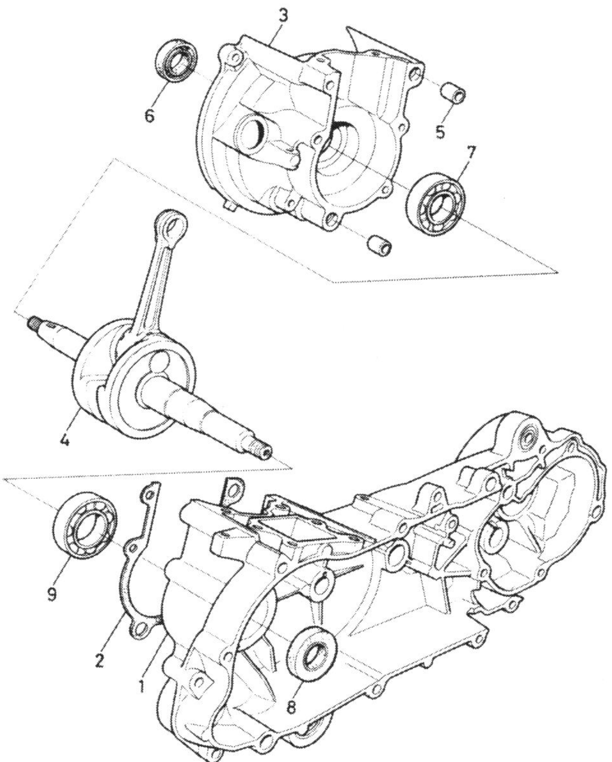

Fig. 1.4 Crankshaft and crankcases

1 *Left-hand crankcase half*
2 *Gasket*
3 *Right-hand crankcase half*
4 *Crankshaft*
5 *Locating dowel – 2 off*
6 *Right-hand oil seal*
7 *Right-hand main bearing*
8 *Left-hand oil seal*
9 *Left-hand main bearing*

18 Examination and renovation: crankshaft assembly

1 The crankshaft assembly comprises two full flywheels, two mainshafts, a crankpin and big-end bearing, a connecting rod and a caged needle roller small-end bearing. The general condition of the big-end bearing may be established with the assembly removed from the engine, or with just the cylinder head and barrel removed, as would be the case during a normal decoke. In this way it is possible to decide whether big-end renewal is necessary, without a great deal of exploratory dismantling.
2 Big-end failure is characterised by a pronounced knock which will be most noticeable when the engine is worked hard. The usual causes of failure are normal wear, or a failure of the lubrication supply. In the case of the latter, big-end wear will become apparent very suddenly, and will rapidly worsen. Check

for wear with the crankshaft set in the TDC (top dead centre) position, by pushing and pulling the connecting rod. No discernible movement will be evident in an unworn bearing, but care must be taken not to confuse end float, which is normal, and bearing wear.

3 If the crankshaft assembly is to be examined with it removed from the engine, then the degree of wear in the big-end can be assessed with a greater amount of accuracy if the assembly is first washed in fuel to remove any residual oil.

4 The big-end bearing is, like the small-end bearing, of the caged needle roller type and under normal circumstances will give good service. If a dial gauge is readily available, an accurate measurement of radial wear in the big-end bearing may be made by positioning the gauge pointer at the positions indicated in the figure accompanying this text. If the measurement taken exceeds the service limit of 0.05 mm (0.002 in), then the bearing must be renewed.

5 Measure the big-end side clearance by inserting the blade of a feeler gauge between the side of the connecting rod at the big-end eye and the crankshaft flywheel. If the clearance measured exceeds the service limit of 0.60 mm (0.024 in), then the bearing must be renewed.

6 Measurement of crankshaft deflection should only be made with the crankshaft assembly removed from the crankcase and set up on V-blocks which themselves have been positioned on a completely flat surface. The amount of deflection present in each mainshaft should not exceed the service limit of 0.10 mm (0.004 in) in the left-hand shaft or 0.15 mm (0.006 in) in the right-hand shaft. Each measurement should be made at the point on the shaft indicated in the figure accompanying this text. If either one of these service limits is exceeded, then the complete crankshaft assembly must be replaced with a serviceable item.

7 Like the big-end bearing, the small-end bearing should seldom give trouble unless a lubrication failure has occurred. Wash the small-end bearing thoroughly in fuel and then push it into the small-end eye of the connecting rod. Push the gudgeon pin through the bearing. Hold the connecting rod steady and feel for any discernible movement between it and the gudgeon pin. If movement is detected, do not automatically assume that the bearing is worn but check that the bore of the small-end eye and the outer diameter of the gudgeon pin are not worn beyond the service limits given in the Specifications Section of this Chapter. If both the connecting rod and gudgeon pin are found to be within limits, discard the bearing and replace it with a new item. Close inspection of the bearing will show if the roller cage is beginning to crack or wear, in which case the bearing must be renewed, even though play may be within the given limits.

8 If any fault is found or suspected in any one of the component parts which make up the crankshaft assembly, then it is recommended that the complete crankshaft assembly is taken to a Honda service agent, who will be able to confirm the worst and supply a new or service-exchange assembly. The big-end bearing is reached by dismantling the pressed-up flywheel assembly and this must be considered to be a job for a specialist. On completion of bearing renewal, the assembly must be trued up and balanced, a job requiring skills and facilities beyond the reach of most home mechanics. Much of the cost of reconditioning the big-end assembly will have been saved by removing the engine and separating the crankcases.

9 Finally, do not omit to check the condition of the worm gear which forms part of the right-hand mainshaft. If this gear is seen to be broken or badly worn, then the right-hand mainshaft will have to be renewed.

18.9 Inspect the crankshaft worm gear

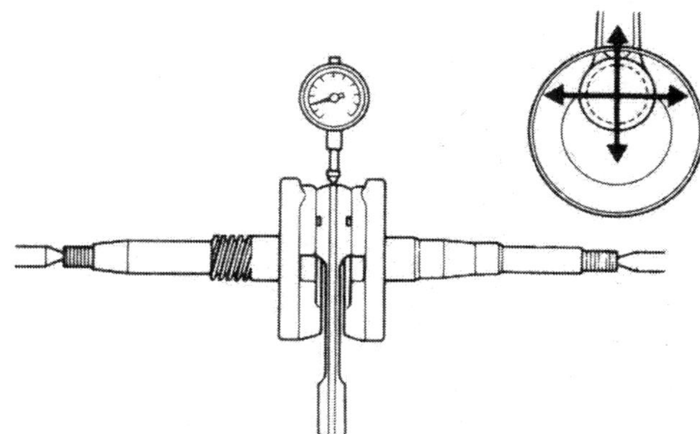

Fig. 1.5 Big-end bearing radial wear measurement

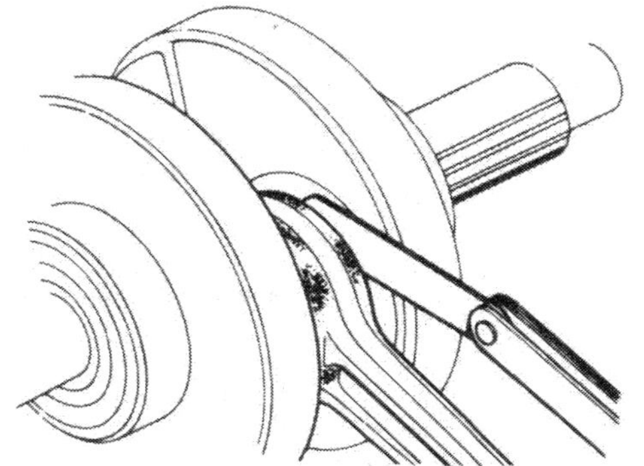

Fig. 1.6 Big-end bearing side clearance measurement

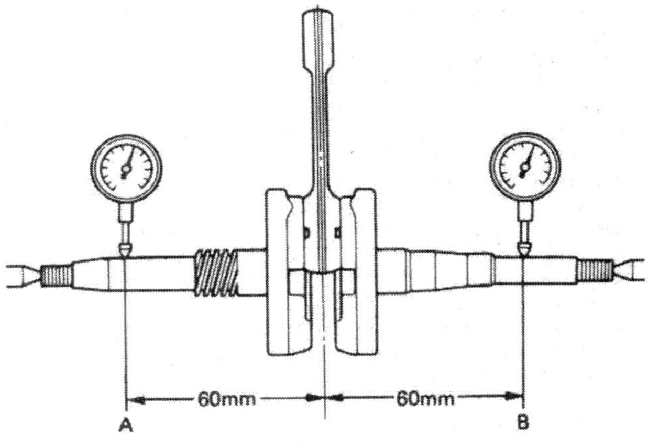

Fig. 1.7 Measuring the amount of crankshaft deflection

19 Examination and renovation: main bearings, transmission bearings and oil seals

1 The main bearings should be washed out petrol and checked for free play. A very small amount of movement is permissible, but severe axial play or any roughness when the bearing is rotated indicates the need for renewal.

2 In practice, it was found that both main bearings remained in their respective crankcase halves instead of being driven out of position with the crankshaft. If this is the case, then both bearings can be driven out of their location in the crankcase once their oil seals have been removed. Make a careful note of the fitted position of each seal. Remove each seal by prising it out of position with the flat of a screwdriver. Take great care to avoid damaging the alloy in or around the seal housing.

3 With the crankcase castings properly supported, drive each main bearing from position by using a length of tube of suitable diameter and a hammer. Warming the castings will greatly aid removal of the bearings. Take care not to overdo this. Only a gentle, generalised form of heat should be applied as the application of a localised, fierce heat, such as the flame of a blowlamp, will only result in the castings becoming distorted. Placing each casting in a container and then immersing it in boiling water is the method recommended.

4 Where the main bearings have remained attached to the crankshaft, it is recommended that the crankshaft assembly is returned to a Honda service agent who will have the proper equipment necessary to draw each bearing off its shaft. If the services of an agent are not available, then it is recommended that the following method be used.

5 Insert two thin steel wedges, one on each side of the bearing, and with these clamped in a vice hit the protected end of the crankshaft squarely with a soft-faced hammer in an attempt to drive the crankshaft through the bearing. When the bearing has moved the initial amount, it should be possible to insert a conventional two or three-legged sprocket puller, to complete the drawing-off action. Great care should be taken when using this technique.

6 The bearing located in the cover of the reduction gearbox and the stub-axle bearing located in the transmission casing can both be removed in a manner similar to that described in the previous paragraphs of this Section. The bearing of the gearbox input shaft is, however, fitted into a blind recess in the transmission casing. The only practical method of removing this bearing is to use a slide hammer as shown in the figure accompanying this text.

7 New bearings should be driven into their locations by using a tubular drift, the end of which must be placed against the bearing outer race. Check that the mating surfaces of both the bearing and the casting are clean and ensure that the bearing enters its location squarely to prevent tying or jamming.

8 Each oil seal should be renewed as a matter of course, even if it appears unscathed after removal. Use a method similar to that described for fitting the bearings when fitting each seal and take care to position each seal in its previously noted position before finally driving it home.

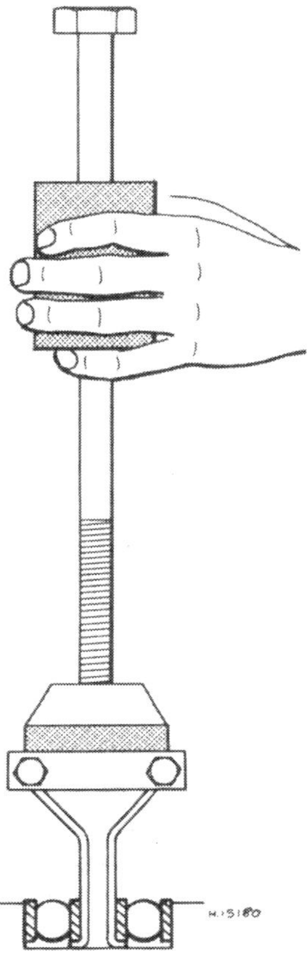

Fig. 1.8 Use of slide hammer to withdraw bearing from blind bearing housing

19.2 Examine the main bearings

19.8 Renew each oil seal as a matter of course

20 Examination and renovation: decarbonisation

1 Decarbonisation must take place as part of any major overhaul, in addition to being a normal routine maintenance function. In the case of the latter, the operation can be undertaken with minimal dismantling, namely removal of the cylinder head, cylinder barrel and exhaust system. Carbon build-up in a two-stroke engine is more rapid than that of its four-stroke counterpart, due to the oily nature of the combustion mixture. It is however, rather softer and is therefore more easily removed.
2 The object of the exercise is to remove all traces of carbon whilst avoiding the removal of the metal surface on which it is deposited. It follows that care must be taken when dealing with the relatively soft alloy cylinder head and piston. Never use a steel scraper or screwdriver for carbon removal. A hard wood, brass or aluminium scraper is the ideal tool as these are harder than the carbon, but no harder than the underlying metal. Once the bulk of the carbon has been removed, a brass wire brush of the type used to clean suede shoes can be used to good effect.
3 The whole of the combustion chamber should be cleaned, as should the piston crown. It is recommended that as smooth a finish as possible is obtained, as this will slow the subsequent build up of carbon. If desired, metal polish can be used to obtain a smooth surface. The exhaust port must also be cleaned out, as a build up of carbon in this area will restrict the flow of exhaust gases from the cylinder. Take care to remove all traces of debris from the cylinder and ports, prior to reassembly by washing the components thoroughly in fuel whilst taking care to observe the necessary fire precautions.
4 Full details of decarbonising the exhaust system are contained in Section 15 of the following Chapter.

21 Examination and renovation: cylinder head

1 Check that the fins of the cylinder head are not clogged with oil or dirt. If heavy contamination is present, then it is likely that the engine will overheat. If necessary, use a degreasing agent and brush to clean between the fins. Check that no cracks are evident, especially in the vicinity of the holes through which the holding down bolts pass, and near the spark plug threads.
2 Check the condition of the thread in the spark plug hole. If it is damaged an effective repair can be made using a Helicoil thread insert. This service is available from most Honda Service Agents. The cause of a damaged thread can usually be traced to overtightening of the plug or using a plug of too long a reach. Always use the correct plug and do not overtighten.
3 If leakage problems have been experienced between the cylinder head and cylinder barrel mating surfaces, the cylinder head should be checked for distortion by placing a straight-edge across several points on its mating surface and then attempting to slide a 0.10 mm (0.004 in) feeler gauge between the straight-edge and the mating surface.
4 If the cylinder head is found to be warped beyond this limit, grind it flat by placing a sheet of emery paper on a surface plate or sheet of plate glass and rubbing the cylinder head mating surface against it, in a slow, circular motion. Commence with 200 grade paper and finish with 400 grade paper and oil. If it is found necessary to remove substantial amount of metal before the mating surfaces becomes completely flat, obtain advice from a Honda service agent as to whether it is necessary to obtain a new head.
5 Note that most cases of cylinder head distortion can be traced to unequal tensioning of the cylinder head securing bolts and to tightening of the bolts in the incorrect sequence.

22 Examination and renovation: cylinder barrel

1 The usual indication of a badly worn cylinder barrel and piston is known as piston slap, a metallic rattle that occurs when there is little or no load on the engine.
2 Commence by cleaning the outside of the cylinder barrel, taking care to remove any accumulation of dirt from between the cooling fins. Carefully remove the ring of carbon from the mouth of the bore, so that an accurate assessment of bore wear can be made.
3 A close visual examination of the bore surface must be made, to check for scoring or any other damage, particularly if broken piston rings were encountered during the stripdown. Any damage of this nature will necessitate renewal of the barrel, as it is impossible to obtain a satisfactory seal if the bore is not perfectly finished.
4 There will probably be a lip at the uppermost end of the cylinder bore which marks the limit of travel of the top of the piston ring. The depth of the lip will give some indication of the amount of bore wear that has taken place even though the amount of wear is not evenly distributed.
5 The most accurate method of measuring bore wear is by the use of a cylinder bore DTI (dial test indicator) or a bore micrometer. Measurements should be made at the top of the bore (just below the wear lip), at a point in the bore just above the ports and at a point just above the cutout in the bore spigot. Make measurements in both a fore and aft and a side to side plane at these positions. If the largest measurement taken exceeds the service limit of 40.050 mm (1.5768 in), then the barrel must be renewed as no rebore sizes are given and no oversize pistons or rings are available. Note that the barrel will have the letter A or B stamped on its crankcase mating surface. The replacement barrel must carry the same letter as the item it is to replace.
6 It is however, most unlikely that the average owner will have the above mentioned items of equipment readily available. A slightly less accurate but more practical method of measurement is as follows. It is possible to determine the amount of wear in the bore by inserting the piston, devoid of rings, into the bore, so that its crown is positioned just below the wear lip at the top of the bore. Measure the distance between the wall of the bore and the top side of the piston by using feeler gauges.
7 Move the piston down the bore so that its crown is positioned in line with a point just above the cutout in the bore spigot. Repeat the measurement. Doing this, and subtracting the lesser measurement from the greater, will give the difference between the bore diameter in an area where the greatest amount of wear is likely to occur and an area in which there should be little or no wear. If the difference found exceeds 0.050 mm (0.002 in), then the barrel will have to be renewed. Bear in mind that the curvature of the gap being measured will tend to preclude accurate measurement by this method but it should be possible to gain a good indication of whether renewal of the barrel is necessary. In view of the cost involved in purchasing a new barrel, it is well having one's worst suspicions confirmed by asking a competent engineer to make an accurate measurement of the bore diameter.
8 Upon receiving a new cylinder barrel, check to confirm that the edges of the ports, where they meet the surface of the bore, have been given a slight chamfer. This chamfer is necessary to prevent the possibility of the piston rings catching on the unchamfered edge of a port and breaking, thus necessitating a further strip down and barrel renewal. Chamfering of the port edge can be carried out by very careful use of a metal scraper but it is essential to ensure that the wall of the bore does not become damaged in the process. Finish off any chamfer made by polishing its cut edges with fine emery paper.
9 Establishment of the clearance between the cylinder bore and the piston can be made either by direct measurement of the cylinder bore and piston diameters, and then by subtracting the latter figure from the former, or by actual measurement of the

gap using a feeler gauge. In either case, if the clearance exceeds the maximum wear limit of 0.10 mm (0.004 in), then there is evidence that either a new piston or a new cylinder barrel is required.
10 If the method of direct measurement of the piston and cylinder bore is decided upon, then the measurement of piston diameter should be made at a point 4 mm (0.16 in) from the base of the piston skirt, at right-angles to the gudgeon pin hole. Full details of the method of measuring bore wear are given in paragraph 5 of this Section.
11 Finally, check the cylinder barrel to cylinder head mating surface for distortion by placing a straight-edge across several places on it and attempting to slide a 0.10 mm (0.004 in) feeler gauge between the straight-edge and mating surface. If the cylinder barrel proves to be warped beyond this limit, grind it flat by placing a sheet of emery paper on a surface plate or sheet of plate glass and rubbing the mating surface against it, in a slow, circular motion. Commence this operation with 200 grade paper and finish with 400 grade paper and oil. If it is thought necessary to remove a substantial amount of metal in order to bring the mating surface back to within limits, obtain advice from a Honda service agent as to whether it is necessary to obtain a replacement barrel.

23 Examination and renovation: piston and piston rings

1 Remove each ring from the piston by pushing its ends apart with the thumb of each hand before gently easing it out of its groove. Great care is necessary throughout this operation because the rings are brittle and will break easily if over-stressed. If the rings are gummed in their grooves, three strips of tin can be used, to ease them free, as shown in the accompanying illustration. Take great care to keep the rings separate and the right way up so that they may be refitted in their correct positions.
2 Piston wear usually occurs at the skirt or lower end of the piston and takes the form of vertical streaks or score marks on the thrust side. There may also be some variation in the thickness of the skirt. Measurement of the piston diameter should be made at a point 4 mm (0.16 in) from the base of the piston skirt, at right-angles to the gudgeon pin hole. If the measurement obtained is found to be less than the service limit of 39.90 mm (1.5709 in), then the piston must be renewed.
3 Check that the piston and cylinder bore are not scored, particularly if the engine has tightened up or seized. If the bore is badly scored, then a new barrel will be required as no rebore sizes are given and no oversize pistons or rings are available. If scoring of the piston is not too severe or the piston has just picked up, it is possible to remove the high spots by careful use of a fine swiss file. Application of chalk to the file teeth will help prevent clogging of the teeth and the subsequent risk of scoring.
4 Check for any build up of carbon in the piston ring grooves. Any carbon present should be carefully removed by using a section of broken piston ring or similar. The grooves in which the piston rings locate can become enlarged in use. Unfortunately, no figure is published for the maximum side clearance. If, however, the clearance exceeds about 0.1 mm (0.004 in) with an unworn ring fitted, ring float may lead to early ring breakage. Before renewing the piston seek the advice of a Honda service agent.
5 When cleaning the piston ring grooves of carbon, check that each ring locating peg is not loose or worn. If in doubt as to the condition of these pegs, then seek professional advice from a Honda service agent and renew the piston if necessary. It is imperative that these locating pegs should not work loose, otherwise the rings will be free to rotate and there is a danger of the ends being trapped in the ports.
6 It cannot be over-emphasised that the condition of the piston and piston rings is of prime importance because they control the opening and closing of the ports by providing an effective moving seal. A two-stroke engine has only three working parts, of which the piston is one. It follows that the efficiency of the engine is very dependent on the condition of the piston and the parts with which it is closely associated.
7 Examine the gudgeon pin for any scoring on its bearing surface. If damage is apparent on its contact surface with the holes in the piston, examine the surface of these holes and reject the piston if similar damage is found. The gudgeon pin should be a firm press fit in the piston. If any excessive looseness is felt between the two components, check the pin to piston clearance and compare the reading obtained with the service limit of 0.040 mm (0.0016 in). If this reading is found to be excessive, measure the gudgeon pin outer diameter and the piston hole bore diameter before rejecting one or both of the components, as necessary.
8 Check that the gudgeon pin circlip retaining grooves in the piston are free from damage. If there is the slightest doubt as to the condition of these grooves, seek further advice from a Honda service agent who will determine whether or not the piston should be renewed. Remember that if one of these rings should become detached in use, certain damage will be caused to both the piston and the surface of the bore.
9 Examine the working surface of each piston ring. If discoloured areas are evident, the ring should be renewed because these areas indicate the blow-by of gas.
10 Piston ring wear is measured by inserting each ring in a part of the cylinder bore which is not normally subject to wear. Ideally, the ring should be inserted into the cylinder bore so that it is positioned approximately 10 mm (0.4 in) from the bore spigot cutouts. Use the crown of the piston as a means of locating the ring squarely in the bore and measure the gap between the ring ends with a feeler gauge. If the gap measured is found to be greater than the service limit of 0.60 mm (0.024 in), then the ring must be renewed. Note that the top ring has the identification mark 1T on its upper surface whereas the second ring has the identification mark 2T. The two rings should never be interchanged as they have different cross-sectional shapes. Always renew the rings as a set; this includes the expander ring which is fitted beneath the second ring.
11 It must be appreciated that, if a set of new rings is to be fitted to a used piston which is to run in a part worn bore, then the upper of the two rings should be stepped to clear the ridge left in the bore by the previous ring. If a new ring is fitted then it will hit the ridge and break. This is because the new ring will not have worn in the same way as the old, which will have worn in unison with the ridge.
12 Honda do not supply a stepped upper ring to fit the Melody. It is necessary, therefore, to ensure that the wear ridge at the top of the bore is completely removed before fitting a set of new rings. This may be incorporated in the 'glaze busting' process, which should be carried out in order to break down the surface glaze on the used bore so that the new rings are able to bed in to the bore surface. This combined process of ridge removal and 'glaze busting' can be carried out by any one of many competent engineering firms who advertise regularly in the national motorcycle press; your local Honda service agent could well offer a similar service.
13 Bear in mind, when refitting used rings to the piston, that they must be fitted in exactly the same positions as noted during removal. When fitting either new or old rings, take note of the following points. Each ring should be fitted to the piston by pulling its ends apart just enough to allow it to pass over the piston crown and into its groove. Take care to fit both the top and second rings with their identification marks facing the piston crown. Ensure that the expander ring has its ends located one each side of the ring locating peg before fitting the second ring into its groove. Each ring must press easily into its groove and have its ends located correctly over the locating peg.
14 Finally, do not automatically assume when fitting new rings, that their end gaps will be correct. As with part worn rings, the end gap must be measured. It may be necessary to enlarge the gap, in which case this should be done by careful use of a needle file. With the rings fitted to the piston, place the assembly to one side ready for engine reassembly.

23.13 Correctly locate the ring ends over their locating peg

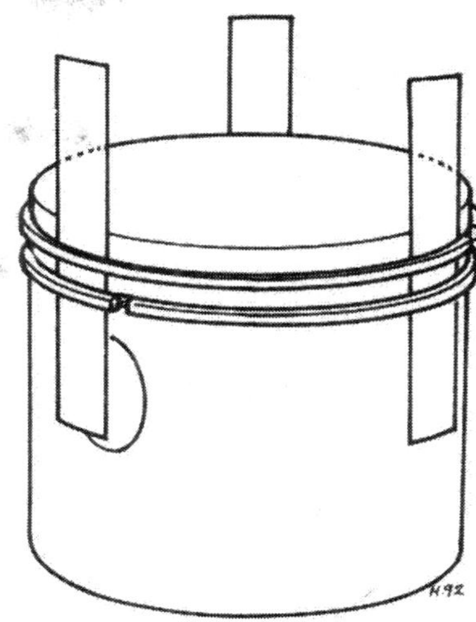

Fig. 1.9 Method of removing gummed piston rings

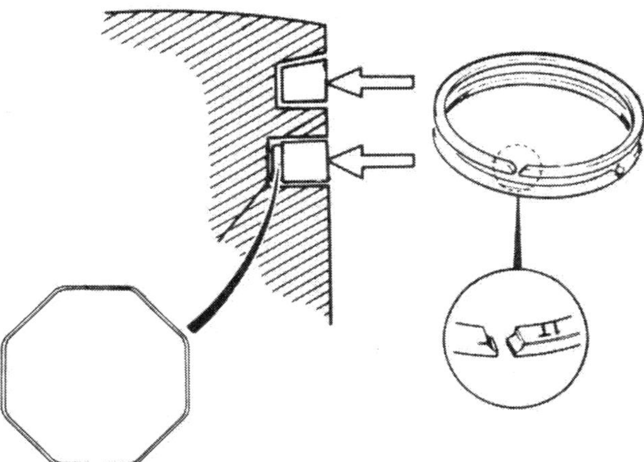

Fig. 1.10 Piston ring profiles

24 Transmission: general description and operation

1 The Honda Melody employs a variable ratio type of transmission system which is designed to provide acceptable machine acceleration in conjunction with good hillclimbing ability.

2 With the engine running at tickover speed, the crankshaft-mounted drive pulley unit transmits power to a driven pulley which is mounted on the input shaft to the reduction gearbox and which forms part of the centrifugal clutch assembly. This it does by means of a toothed V-belt. This driven pulley is designed to rotate independently of the gearbox input shaft at engine tickover speed and therefore no power is transmitted into the reduction gearbox and hence to the rear wheel.

3 Opening of the throttle will increase engine speed; as this happens, the spring-loaded shoes attached to the backplate of the centrifugal clutch assembly will move from their restrained position against their stops and be thrown outwards by the increased centrifugal force imposed upon them until they begin to come into contact with the clutch drum. From this point, where the shoes are merely rubbing on the drum surface, any further increase in engine speed will lead to greater pressure being applied by the shoes to the drum until the drive between the clutch backplate and the clutch drum is solid. Because the clutch backplate is locked to the input shaft of the reduction gearbox, this means that a smooth take up of power is now being transmitted to the rear wheel, therefore causing the machine to move forward.

4 The road speed which the machine will now attain is directly dependent on the engine speed and the gearing ratio between the drive and driven pulleys. This system does, however, have its limitations; a good example being when the machine is required to climb a fairly long and steep hill. As the slope of the hill begins to take effect on the speed of the machine, the only way that the rider can compensate for this decrease in speed is to open the throttle. It then becomes obvious that once the throttle is fully open, nothing more can be done to prevent the machine from slowing down and eventually stopping. Honda's answer to this problem is to provide a method of varying the gearing ratio between the drive and driven pulleys. The method used to achieve this is explained in the following paragraphs.

5 Both of the pulley units incorporated in the transmission of the Honda Melody are of the variable diameter type. With the engine at rest, the two faces of the drive pulley are at their furthest point apart, therefore making the effective diameter of the pulley as small as possible. The sliding (outboard) face of the driven pulley is spring loaded so that it is pushed hard against the fixed face of the pulley whilst at rest, thereby giving the largest possible pulley diameter. This combination of pulley sizes gives a low overall gearing which helps to provide good acceleration of the machine from rest.

6 As the engine speed is increased, the sleeved rollers incorporated within the sliding (inboard) face of the drive pulley are forced outwards by the centrifugal force imposed upon them. This causes the sliding face of the pulley to move outwards, thereby increasing the effective diameter of the pulley. Directly this happens, the increased tension on the V-belt causes the sliding face of the driven pulley too be forced outwards against its spring, thereby reducing the effective diameter of the pulley and at the same time allowing the drive belt to return to its correct tension. The end result of this is that the overall gearing is raised, therefore giving a higher road speed for a lower engine speed.

7 Directly the engine speed is allowed to fall, the reduction in the amount of centrifugal force imposed upon the sleeved rollers of the drive pulley unit will allow them to return inwards. As the faces of the drive pulley move apart and the tension on the drive belt decreases, the spring loading on the sliding face of the driven pulley will cause it to be forced hard against the fixed face of the pulley. This increases the effective diameter of the driven pulley, which in turn places enough tension on the drive belt to ensure that it keeps the sliding face of the drive pulley pressed hard against the receding rollers. The overall gearing is, therefore, effectively lowered.

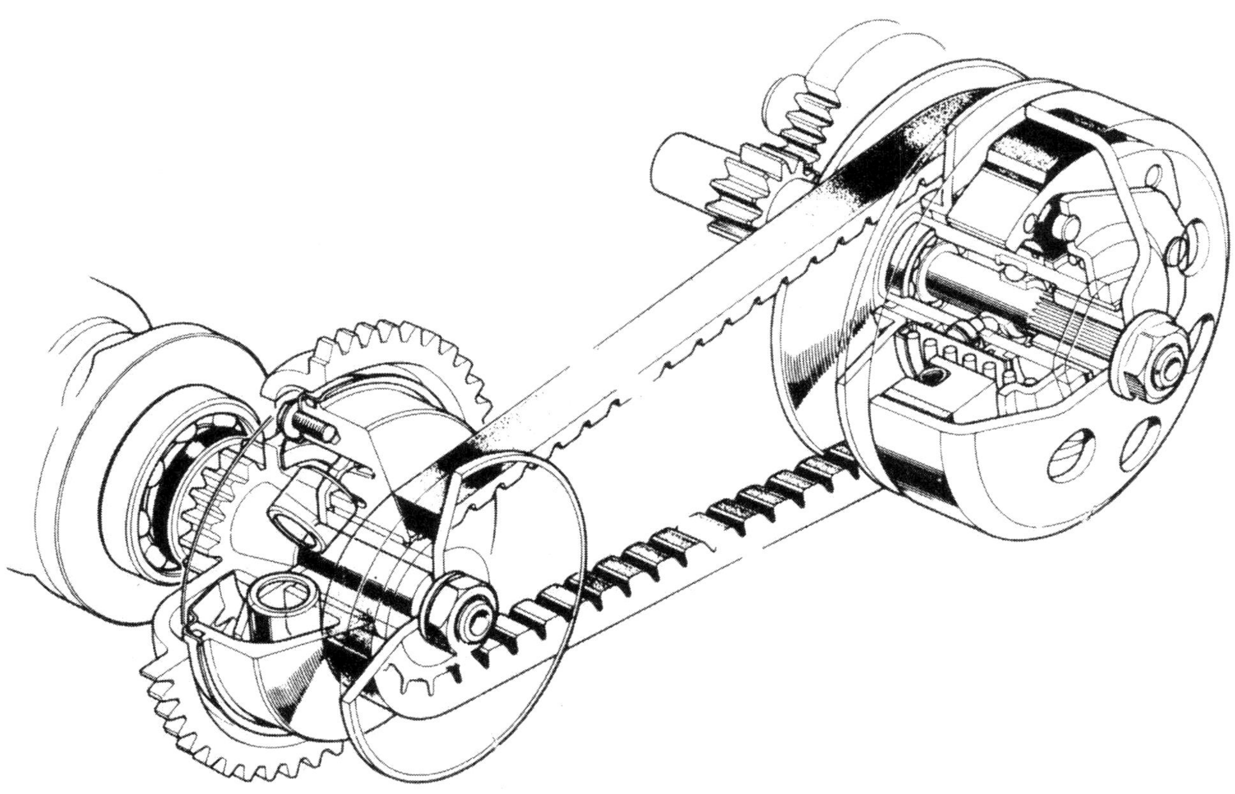

Fig. 1.11 Sectional view of drive pulley arrangement

25 Examination and renovation: drive pulley unit

1 Place the sliding face assembly of the drive pulley unit on a clean work surface and remove the three bolts which retain the cover plate in position. Removal of this plate will reveal the ramp plate located beneath it. Ease this plate out of the face housing and detach each one of the three nylon slide pieces from its outer edge. Place the complete assembly to one side, ready for inspection.

2 Lift each one of the six rollers out of the face housing and wipe them clean before placing them to one side, ready for inspection. Wipe any grease out of the housing and detach the O-ring from its outer edge. Push the drive face boss out of the centre of the housing.

3 The face housing should now be closely examined for signs of fatigue or failure, such as the presence of hairline cracks around the base of the webs within the housing or around the threaded bosses of the cover plate retaining bolt locations.

4 Degrease the surface of the bush fitted into the centre of the face housing and degrease the outer surface of the drive face boss. Refit the boss into the bush and check for movement between the two components. If excessive movement is found, one or both of the components will have to be renewed. Measure the overall diameter of the boss. If the measurement taken is less than the service limit of 24.960 mm (0.9827 in) then the boss must be renewed. The internal diameter of the bush should not be more than the service limit of 25.130 mm (0.9894 in). If the bush is worn beyond this limit then it must be renewed. Honda do not, however, list this bush as a separate item, which indicates that if the bush is unserviceable then the complete housing should be renewed. Before going to the expense of purchasing a new housing, it may be worth approaching a competent engineer who may agree to rebush the housing for less than it would cost to renew it. Examine the condition of the O-ring detached from the housing. If the ring is flattened or in any way damaged, then it must be renewed. Fit the serviceable O-ring to the housing.

5 Carefully examine the bearing surface of each roller sleeve. If any one sleeve is scored or shows signs of having developed a flat spot, then all six sleeves should be renewed. The same applies if the overall diameter of any one sleeve is found to be less than the service limit of 15.40 mm (0.606 in) or if a sleeve is found to be anything other than a firm fit on its roller.

6 Examine the nylon slide pieces removed from the ramp plate for signs of damage or wear. Each piece should be a firm fit on the ramp plate and a close sliding fit over its retaining web in the face housing. If any one piece is found to be defective, then all three pieces should be renewed. Examine the ramp plate for signs of fatigue or failure before fitting the serviceable slide pieces to its outer edge. Finally, examine the cover plate for signs of fatigue.

7 Reassemble the drive pulley unit by reversing the dismantling procedure. Lubricate each roller ramp before inserting the rollers into the face housing. With the rollers all fitted, smear the surface of the centre bush and the inside surface of the ramp plate with grease before pushing the plate into position. The grease used should be of a lithium based type. Check, before fitting and tightening the cover plate retaining bolts, that the plate is pushed fully down over the O-ring and that the O-ring remains in its location. These bolts should be tightened to a torque loading of 2 - 3 lbf ft (0.25 - 0.40 kgf m).

25.2 Withdraw the drive face boss

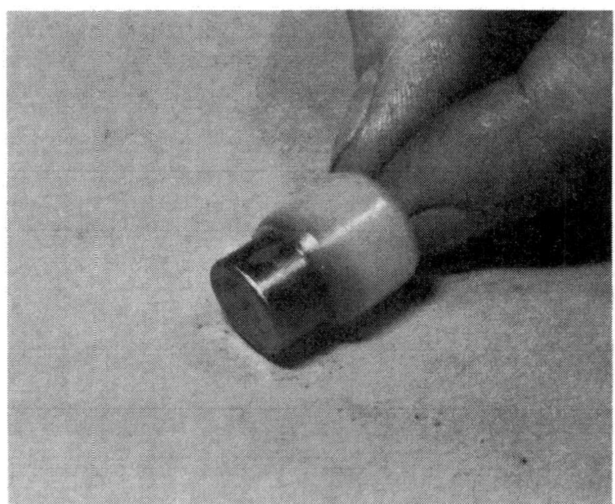
25.5 Each sleeve should be a firm fit over its roller

25.6 Examine the nylon slide pieces

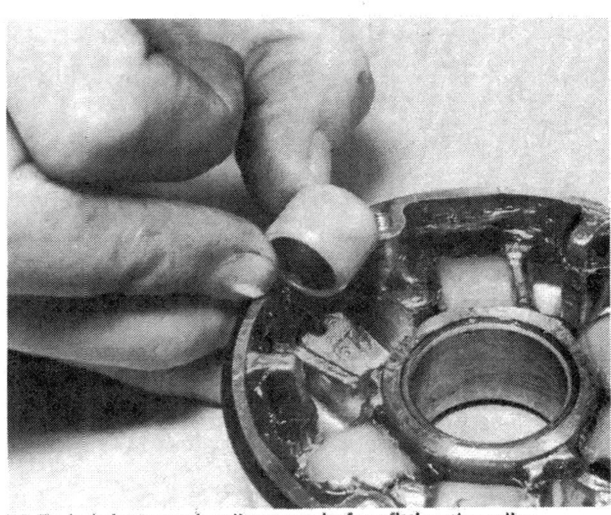
25.7a Lubricate each roller ramp before fitting the roller

25.7b Push the ramp plate into position

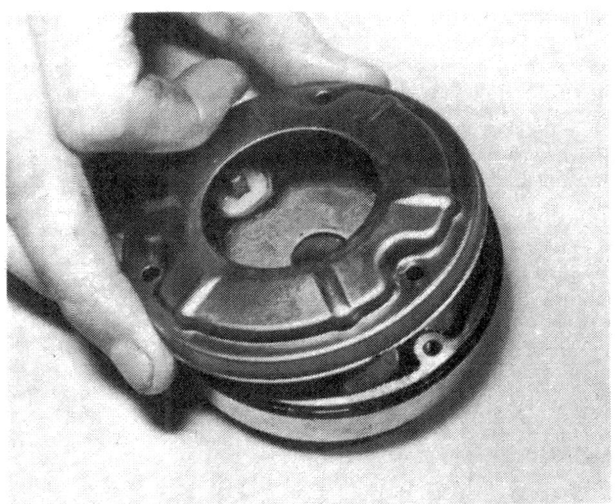
25.7c Fit the cover plate over the O-ring

26 Examination and renovation: drive belt

1 Carry out an examination of the drive belt for signs of cracking on any of its surfaces. Check also that the laminations which make up the belt have not begun to separate. The belt must be renewed if it shows any sign of deterioration. Determine the amount of wear present in the drive belt by taking a measurement of its width across its widest surface. The measurement obtained should not be less than the service limit of 12.5 mm (0.492 in). If the belt is worn beyond this limit then the efficiency of the transmission system will be adversely affected and the belt will fail before many more miles are covered with the machine.

Fig. 1.12 Drive belt wear measurement points

27 Examination and renovation: centrifugal clutch and driven pulley unit

1 The centrifugal clutch and the driven pulley effectively comprise a single unit which is mounted on the input shaft to the reduction gearbox. If it is found that the engine speed is required to be abnormally high before power is transmitted to the rear wheel, then the clutch unit must be removed from the machine and the shoe linings inspected for excessive wear, scoring, or contamination by oil or grease. If any of these factors are evident, then the shoes will require renewal.
2 If it is found that the machine has a tendency to creep forwards when the engine is running at tickover speed, the clutch shoe return springs may have weakened or stretched thus allowing the shoes to come into engagement prematurely with the clutch drum. If this fault is suspected, then the clutch unit must be removed from the machine and the springs detached from the unit so that they may be inspected and, if necessary, renewed.
3 Place the clutch unit on a clean work surface ready for dismantling. Note, before attempting to dismantle the unit, that it contains a large spring which is held in compression between the sliding face of the driven pulley and the clutch backplate. It follows, therefore, that considerable care must be taken whilst removing the backplate retaining nut to prevent the force of the spring from stripping the holding threads of the retaining nut as it is being removed from the boss of the fixed face of the pulley and causing the backplate to be ejected upwards into the face of the person removing the nut.
4 Honda supply a set of special tools which may be used to hold the clutch backplate against the pulley unit whilst the retaining nut is removed and then allow the backplate to be released from the pulley unit under controlled conditions. The tools which make up this set are as follows:

Clutch spring compressor holder	07960 - GA70100
Locknut spanner	07916 - 1870001
Clutch spring compressor	07960 - 0110000
Spring compressor attachment	07960 - 1870100

If the above listed set of tools cannot be easily borrowed or hired, then it is possible to dismantle the unit without them but this will require some care and the help of an assistant. Proceed as follows.
5 With the unit placed on the work surface so that the backplate retaining nut is facing upwards, place the heel of each hand on the surface of the backplate as shown in the photograph accompanying this text. Push down on the backplate to release spring pressure from the retaining nut and loosen and remove the nut. The backplate can now be slowly eased upwards off its retaining boss and lifted clear of the pulley assembly. Remove the spring and pull the seal collar off the pulley boss. It should be noted that because the backplate retaining nut is of a large width (28 mm) yet very thin in section, some difficulty may be experienced in finding a spanner with which to loosen it. A ring spanner is essential for this task as the nut was found to be fairly tight and had to be gripped positively by the spanner before any force could be imposed upon it. The spanner shown in the photograph accompanying this text is similar to the one supplied by Honda for the purpose but was in fact the spanner supplied for removal of the flywheel retaining nut from a Mini.
6 With the seal collar removed from the pulley boss, the sliding face of the pulley can be detached from the boss of the fixed face by withdrawing each of the two guide pins with their rollers from the slots in the sliding face and then pulling the sliding face from position.
7 Commence examination of the clutch unit by inspecting the clutch drum for distortion. This can be done by placing the drum, outboard side uppermost, on a completely flat surface, such as a sheet of plate glass. Any distortion in the drum will be indicated by a gap between the edge of the drum and the glass. A distorted drum must be replaced with a serviceable item as it will adversely affect the smooth transference of power from the crankshaft to the rear wheel. If any distortion found is not too great, it may be possible to have the drum trued, although it is not possible to move more than a small amount of metal from the inner face of the drum without causing it to become severely weakened. Obtain specialist advice before attempting this task. Note that the service limit given for the internal diameter of the drum is 107.5 mm (4.23 in). This diameter should be measured as a matter of course and the drum rejected if necessary. Inspect the inner face of the drum, where it comes into contact with the clutch shoe linings, for scoring. Any bad scoring of the face will adversely affect the efficiency of the clutch unit. Finally, slide the drum over the end of the input shaft and feel for play between the two components. Any excessive play will mean that one or both sets of splines have worn beyond limits, necessitating either renewal of the shaft or drum, or both. Because of the cost involved in renewing each one of these components, it is well worth referring the problem to a Honda service agent before going ahead with any purchase of items.
8 Examine the lining of each clutch shoe for excessive wear, for scoring or for contamination by oil or grease. The shoe lining must not be worn beyond the service limit of 2.0 mm (0.079 in). If a decision is made to reject any one of the clutch shoes then all three of the shoes will have to be renewed as a complete set. Note that there is no satisfactory way of degreasing the lining material.
9 Each clutch shoe can be removed by first unhooking the ends of the return springs from it and then removing the E-clip with its plain washer from the shoe pivot pin before lifting the shoe clear of the clutch backplate. It was found that some effort was needed to unhook the return springs from each shoe; the method used being to place the flat of a small screwdriver between the spring end and the shoe before using the screwdriver as a lever with which to detach the spring.
10 Carefully inspect each one of the three small rubber rings which act as return dampers for the clutch shoes. These must be renewed if they are seen to be in any way damaged or have deteriorated. Fitting of the new clutch shoes may be achieved

by reversing the procedure used for removing the old items. Before pushing each shoe into position over its pivot pin, ensure that the pin is free of corrosion and its bearing surface very lightly lubricated with a lithium based grease. Once fitted, check that each shoe is free to move around its pivot pin.

11 The clutch shoe return springs should be checked for signs of damage or deterioration before being refitted. Any spring that has become corroded should be renewed as its strength will have been affected by the reduction in material thickness. Place all three springs against one another and compare their lengths. If any one spring has set to a length which is greater than the other two, then all three springs should be renewed. If creeping forwards of the machine has been experienced, then the free length of each spring will have to be compared with that of a new item as Honda do not supply a service limit for spring free length. With the serviceable springs fitted, the clutch assembly should now be placed in clean, safe storage ready for fitting to the pulley unit.

12 Commence examination of the driven pulley unit by measuring the free length of the sliding face return spring. The spring must be renewed if the measurement obtained is less than the service limit of 88.0 mm (3.46 in).

13 Inspect the seal collar for any obvious damage or wear, especially around its contact area with the return spring. Renew the collar if necessary.

14 Clean and degrease the boss of the pulley fixed face and examine it for scoring. Any excessive damage to the surface of the boss will necessitate renewal of the pulley face. Measure the overall diameter of the boss. If the measurement obtained is less than the service limit of 33.930 mm (1.3358 in) then the component must be renewed.

15 Insert each one of the sliding face guide pins into the fixed face boss. Feel for any excessive play between the pins and their retaining holes. Any elongation of the retaining holes will necessitate replacement of the fixed face whereas wear on the pins will be obvious by visual inspection. Check also for wear on the bearing surfaces of each guide roller. Ideally, the two pins with their rollers should be renewed as a set.

16 Rotate each of tbe bearings contained in the boss of the fixed face and feel for any roughness in their movement. Any roughness found will indicate that the bearing concerned is due for renewal. Renewal of a bearing is a straightforward operation which requires the use of a soft-metal drift, a hammer and a length of tube with an outside diameter which is just less than that of the bearing concerned.

17 Commence removal of each bearing by displacing its retaining circlip (where fitted) and then supporting the pulley face on a solid work surface so that there is enough room between the boss and the surface for the bearing to be ejected. Displace the bearing by placing the end of the drift against the inner face of the bearing and then striking the drift with the hammer. Move the end of the drift around the periphery of the bearing to keep the bearing square in its location and thus prevent it from binding.

18 Before fitting the new bearing, clean and degrease both the bore of the boss and the contact face of the bearing. Lubricate the bearing with a high melting point, lithium based grease and press it into the boss as far as it will go whilst taking care to keep it square to the boss. Place the end of the length of tube (a suitable socket will also do) on the bearing and carefully tap the bearing home into its location.

19 The two O-rings which fit over the boss of the pulley sliding face should be renewed as a matter of course. The same applies to the seal contained in each end of the pulley boss. Each one of these seals can be removed by carefully levering it from position with the flat of a screwdriver. Each new seal should be fitted by placing a flat piece of wood across the top of the seal and then gently tapping the wood with a hammer to drift the seal into position. Ensure that the lip of each new seal is facing the centre of the pulley boss.

20 The boss of the pulley sliding face should be examined for scoring. Any excessive damage to the boss will necessitate its renewal. Measure the internal diameter of the boss. If the measurement obtained is more than the service limit of 34.060 mm (1.3409 in) then the component must be renewed. Finally, check the guide slots in the boss for signs of excessive wear or damage.

21 With both the centrifugal clutch and the driven pulley components examined and renovated, reassemble them by reversing the dismantling procedure whilst noting the following points. Lightly grease the contact areas of the guide rollers and the guide pins before pushing the seal collar into position over the two O-rings. Use the sliding face return spring as a means of pushing the seal collar fully home against the pulley face. Finally, note that the recommended torque loading for the clutch backplate retaining nut is 25 - 29 lbf ft (3.5 - 4.0 kgf m).

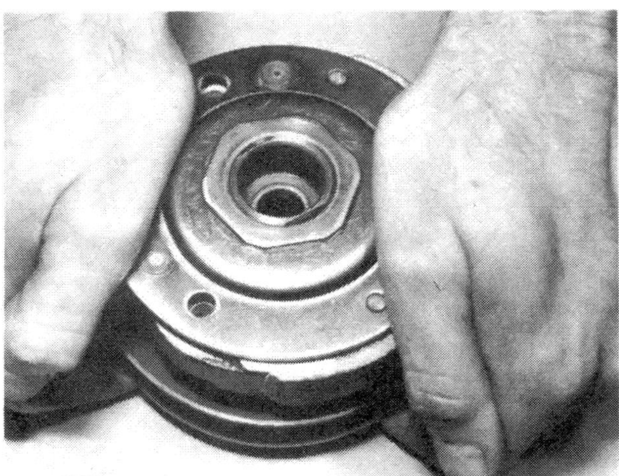

27.5a Push down on the clutch backplate to release spring pressure from the nut

27.5b The spanner used to remove the nut must be a good fit

27.5c Spring pressure will force the unit apart

27.6a Detach the seal collar ...

27.6b ... remove the guide pins and rollers ...

27.6c ... and separate the pulley halves

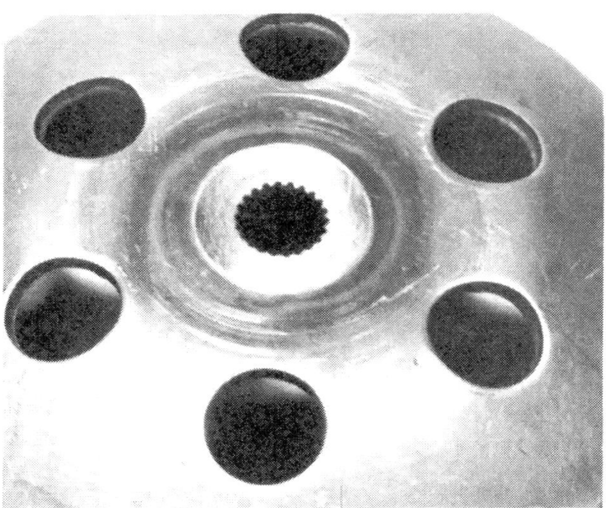
27.7 Examine the clutch drum centre splines

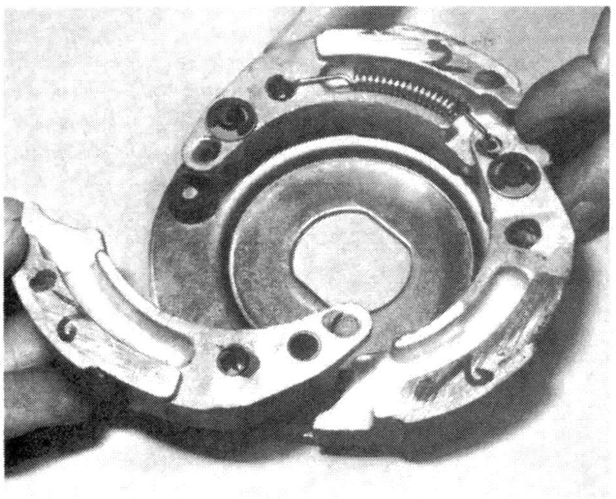

27.9 Detach each clutch shoe ...

27.10 ... and inspect each of the damper rings

27.11a Use a small screwdriver to stretch each spring ...

27.11b ... so that it is correctly fitted between clutch shoes

27.17a Remove each bearing from the fixed face of the pulley

27.17b Note the bearing retaining circlip

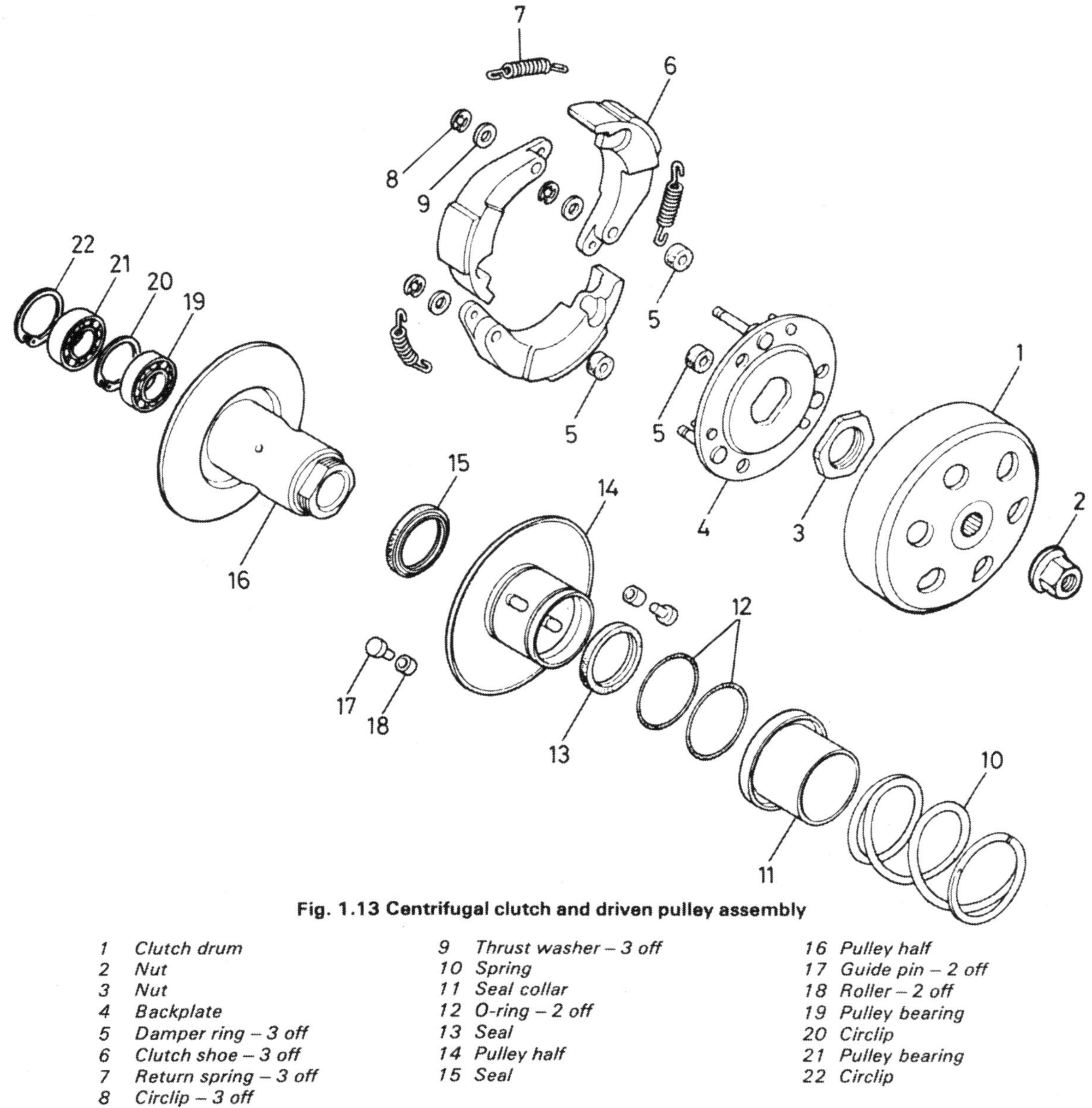

Fig. 1.13 Centrifugal clutch and driven pulley assembly

1 Clutch drum
2 Nut
3 Nut
4 Backplate
5 Damper ring – 3 off
6 Clutch shoe – 3 off
7 Return spring – 3 off
8 Circlip – 3 off
9 Thrust washer – 3 off
10 Spring
11 Seal collar
12 O-ring – 2 off
13 Seal
14 Pulley half
15 Seal
16 Pulley half
17 Guide pin – 2 off
18 Roller – 2 off
19 Pulley bearing
20 Circlip
21 Pulley bearing
22 Circlip

28 Examination and renovation: reduction gearbox components

1 Carefully examine each of the gear pinions to ensure that there are no chipped or broken teeth. Gear pinions with this defect must be renewed as there is no satisfactory method of reclaiming them. Look also for signs of pitting on the load bearing faces of the teeth and for signs of severe wear which may have been caused through lack of lubrication. If damage or wear warrants renewal of any gear pinions, then the complete gearbox assembly should be dismantled and laid out on a clean work surface. The same applies if any one of the gearbox shafts is seen to be worn or damaged.

2 If the decision is made to dismantle the gearbox assembly for the purposes of component renewal and further examination, then each component part should be either renewed, if seen to be obviously damaged, or washed in fuel and placed to one side ready for further examination and reassembly. Remember to observe the necessary fire precautions whilst using fuel to clean the gearbox components.

3 Lay both the new and cleaned components out on a sheet of clean paper or rag and proceed to carry out a further examination of the remaining original components as follows.

4 Fit the intermediate gear pinion to its shaft and feel for any excessive play between the two components. If any play found is thought to be unacceptable, then refer the problem to a Honda service agent who will be able to give an expert opinion and, if necessary, obtain the necessary replacement parts.

5 Check the condition of the splines of each shaft. Closely inspect the load bearing surfaces of each spline for signs of deterioration in its surface finish. This, and any wear lip found along the length of each spline, will necessitate renewal of the shaft concerned.

6 Fit the ends of the intermediate and output shafts into their location in the gearbox cover and check for excessive play between each shaft and its location. Do the same with the intermediate shaft where it fits into the gearbox casing. Again,

if any play found is thought to be unacceptable, then refer the problem to a Honda service agent. It will be seen that neither the casing nor the cover of the gearbox are bushed to accept the shafts, which means that Honda recommend renewal of the complete casting should a shaft location be excessively worn. It may, however, be possible to avoid the expense of purchasing replacement castings by obtaining the advice of a competent engineer as to whether it is possible to have the casting modified to accept bushes.

7 Note that each shaft should be a good sliding fit into its bearing. Any excessive play between the two components will necessitate renewal of the shaft or bearing, or both. Finally, check the condition of the two thrust washers, fitted one each end of the intermediate shaft, and renew them as necessary.

8 Check that each component part, both new and old, is scrupulously clean before lightly lubricating it with clean SAE 10W/30 oil. Reassemble the gearbox component parts to the gearbox cover in the sequence shown by the photographs accompanying this text. Take great care not to allow any grit or moisture to enter between the components whilst doing this and wrap the assembly in a piece of clean rag or paper until it is required for fitting to the transmission casing.

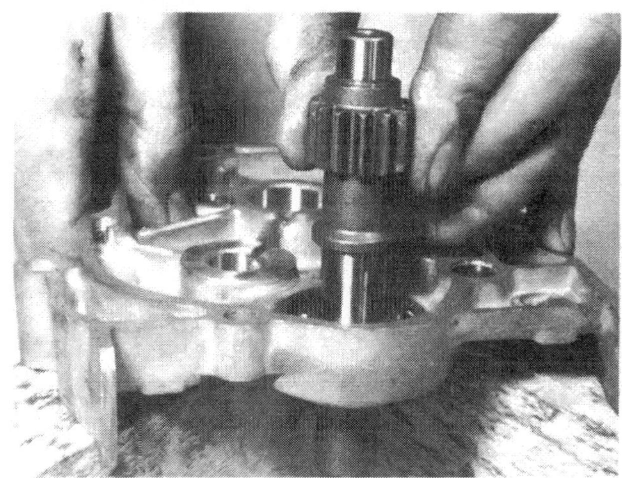

28.8a Insert the input shaft into the gearbox cover ...

28.8b ... followed by the output shaft

28.8c Position the first of the intermediate shaft thrust washers

28.8d ... followed by the shaft itself ...

28.8e ... the intermediate gear pinion and the second thrust washer

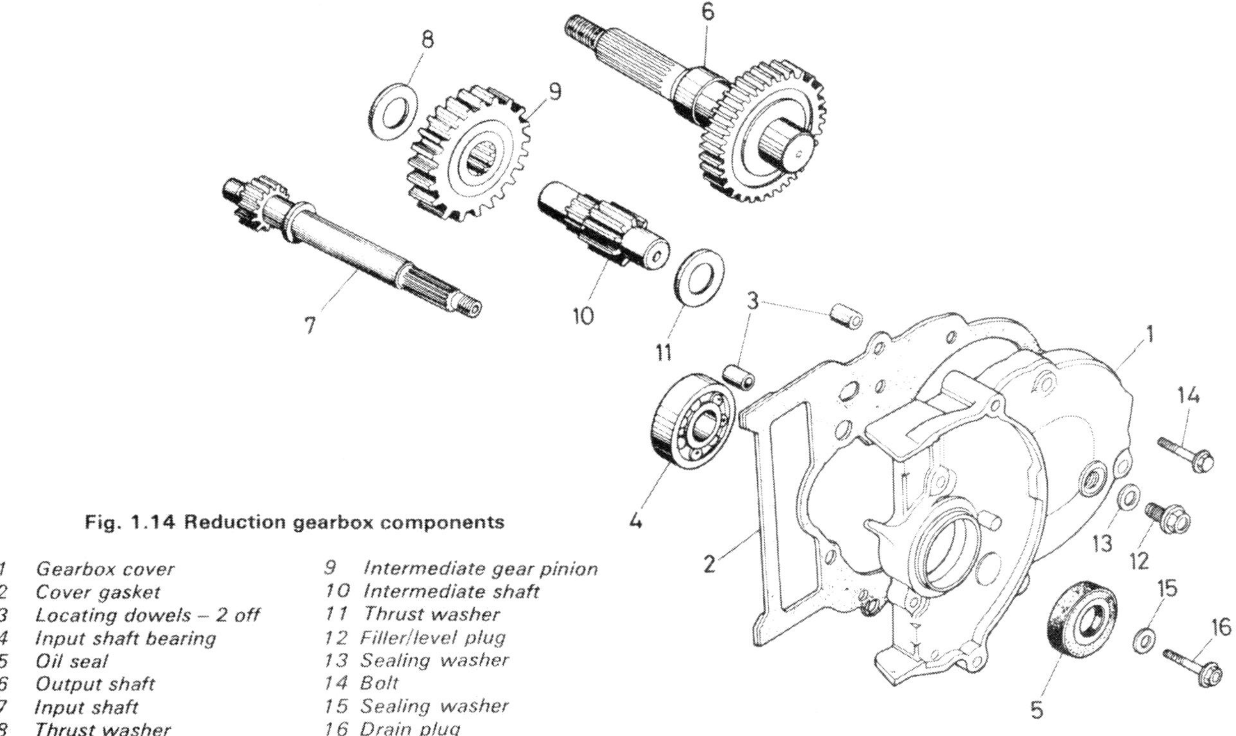

Fig. 1.14 Reduction gearbox components

1 *Gearbox cover*
2 *Cover gasket*
3 *Locating dowels – 2 off*
4 *Input shaft bearing*
5 *Oil seal*
6 *Output shaft*
7 *Input shaft*
8 *Thrust washer*
9 *Intermediate gear pinion*
10 *Intermediate shaft*
11 *Thrust washer*
12 *Filler/level plug*
13 *Sealing washer*
14 *Bolt*
15 *Sealing washer*
16 *Drain plug*

29 Examination and renovation: kickstart components

1 Thoroughly clean each component part of the kickstart assembly in fuel and lay the complete assembly out on a clean work surface in a logical order. Remember to observe the necessary fire precautions whilst cleaning the component parts.
2 Commence examination of the assembly by checking the teeth of each pinion for damage or excessive wear. The meshing teeth will obviously wear in unison, therefore necessitating renewal of the two components concerned. Do not attempt to mate a new component to a part worn item.
3 Inspect the bearing surface of each shaft for signs of heavy scoring. Any damage found will necessitate renewal of the shaft concerned. Select the idle pinion shaft and measure its overall diameter at each end. If the measurements obtained are less than the service limits of 15.90 mm (0.630 in) and 11.90 mm (0.4685 in), reject the shaft and replace it with a new item. Select the kickstart lever shaft and measure its overall diameter at the section of shaft adjacent to the splined end. If the measurement obtained is less than the service limit of 13.90 mm (0.5472 in) then renew the shaft.
4 Insert the kickstart lever shaft through each of its two bushes and check for excessive play between each bush and the shaft. Renew each bush as found necessary. Note that Honda give a service limit for the amount of wear allowed on the internal diameter of the shaft to transmission cover bush as 14.10 mm (0.5551 in).
5 Insert the small pin removed from each shaft back into its location and feel for any excessive play between the pin and the shaft. The pin must be renewed if worn. Any excessive wear in the location holes will, however, necessitate that the shaft concerned be returned to a Honda service agent for further inspection and, if necessary, renewal.
6 It will be seen that the ends of the idle pinion shaft bear directly in the castings of the transmission casing and cover. This means that if any excessive wear is found to exist in either casting bearing surface, then Honda recommend that the complete casting should be renewed. If may, however, be possible to avoid the expense of purchasing replacement castings by obtaining the advice of a competent engineer as to whether it is possible to have the castings modified to accept bushes.
7 Inspect each one of the three springs contained within the assembly for signs of damage or deterioration and renew each one as necessary. No free length service limit is given for the idle pinion spring which will therefore have to be compared with a new item if it is suspected of having taken a set.
8 Check the condition of the single thrust washer which is fitted over the kickstart lever shaft and renew it if necessary. The kickstart return stop should now be examined for deterioration of its rubber buffer or separation of the rubber from the metal plates. A defective stop must be renewed. Check also for excessive wear on the stop retaining plate where it comes into contact with the kickstart return spring and renew the plate as necessary.
9 Finally, on Honda Melody models equipped with a starter motor, inspect the condition of the teeth on the motor pinion. If these are seen to be damaged or excessively worn, then it will be necessary to renew the complete motor pinion assembly. Full details of starter motor renovation are given in Chapter 6 of this Manual.

30 Reassembling the engine/transmission unit: general

1 Before reassembly of the engine/transmission unit is commenced, the various component parts should be cleaned thoroughly and placed on a sheet of clean paper, close to the working area.
2 Make sure all traces of old gaskets have been removed and that the mating surfaces are clean and undamaged. One of the best ways to remove old gasket cement is to apply a rag soaked in methylated spirit. This acts as a solvent and will ensure that the cement is removed without resort to scraping and the consequent risk of damage.
3 Gather together all of the necessary tools and have avail-

able an oil can filled with clean engine oil and one filled with clean gearbox oil. Make sure that all the new gaskets and oil seals are to hand, also all replacement parts required. Nothing is more frustrating than having to stop in the middle of a reassembly sequence because a vital gasket or replacement has been overlooked.

31 Reassembling the engine/transmission unit: joining the crankcase halves

1 Position the transmission casing on a flat, clean work surface so that it is well supported and far enough clear of the surface to allow protrusion of the crankshaft end from the transmission housing. Lubricate the main bearing in the transmission casing and the big-end bearing with clean engine oil and push the crankshaft through the main bearing as far as it will go with hand pressure whilst taking care to keep the crankshaft square to the casing. Fit the generator rotor retaining nut to protect the threaded end of the crankshaft and using a soft-faced hammer, carefully tap the crankshaft home so that it seats against the main bearing. Note that failure to keep the transmission casing well supported will lead to it becoming distorted which will render it effectively useless.

2 Remove the rotor retaining nut from the crankshaft end. Refit the two locating dowels into the crankcase mating surface and position a new crankcase sealing gasket over them. Lubricate with clean engine oil the main bearing contained in the right-hand crankcase half and press the crankcase half down onto the crankshaft as far as it will go with hand pressure. Ensure that the crankcase half is set square to the crankshaft and place two short lengths of wood across its upper mating surface, one each side of the crankshaft. Using a soft-faced hammer, tap down on these lengths of wood to drive the crankcase half home over the locating dowels.

3 Check that both crankcase halves have mated correctly and then fit the six crankcase securing bolts into their previously noted locations. Tighten these bolts evenly and a little at a time whilst working in a diagonal sequence. This will preclude any risk of the crankcase castings becoming distorted as the bolts are tightened.

4 It is now necessary to check that the crankshaft is free to rotate. The crankshaft should spin quite freely if the connecting rod is pushed up and down. If the crankshaft seems reluctant to move, tap it inwards at either end with a soft-faced hammer to free it. Finally, trim the excess gasket from across the crankcase mouth with a sharp knife.

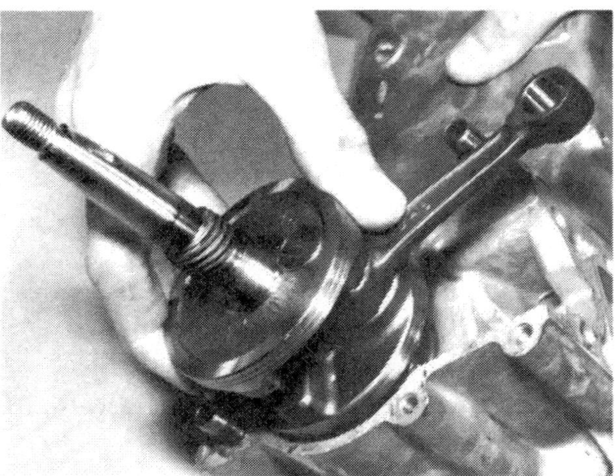
31.1 Insert the crankshaft into the left-hand crankcase half

31.2a Position the crankcase sealing gasket over the locating dowels ...

31.2b ... and press the right-hand crankcase half into position

31.4 Lubricate the big-end bearing

32 Reassembling the engine/transmission unit: fitting the piston, cylinder barrel and cylinder head

1 Position the crankcase unit upright on the work surface and move the connecting rod to its top dead centre (TDC) position. Pack a piece of clean rag between the sides of the connecting rod and the crankcase mouth to prevent any component parts from falling into the crankcase.

2 Fit the small-end bearing into the eye of the connecting rod and lubricate it thoroughly with clean engine oil. Position the piston over the connecting rod so that the 'EX' mark on the piston crown is facing the front of the engine. Push the gudgeon pin into positon. The pin should be a firm sliding fit but if it proves to be tight then warm the piston in hot water to expand the metal around the gudgeon pin bosses.

3 Always use new circlips to retain the gudgeon pin in position and double check that each clip is correctly located in the piston boss groove. Note that each circlip must be fitted so that its gap is well away from the cutouts in the sides of the gudgeon pin hole. A circlip that is allowed to work loose will cause serious damage to both the cylinder bore and piston.

4 Check that the piston rings have not been disturbed from their set positions and then remove the rag from the crankcase mouth. Place a new cylinder barrel base gasket in position on the crankcase. Lubricate the big-end, the piston rings and the cylinder bore with clean engine oil. Grip the piston in one hand whilst pressing the rings into their grooves. The cylinder barrel can now be lifted into position over the crown of the piston and guided carefully down over the piston rings. There is a generous lead-in on the base of the cylinder bore which serves to make this operation reasonably easy. With the piston pushed fully into the bore, align the cylinder barrel with the crankcase mating surface and push it into position.

5 Wipe any excess engine oil off the upper surface of the cylinder barrel and place the new cylinder head gasket in position on the barrel. Fit the cylinder head and push the four retaining bolts into position through the cylinder head and barrel. Do not force these bolts. If difficulty is experienced in fitting them, recheck that the cylinder barrel, cylinder head and their respective gaskets are all in correct alignment with the surface of the crankcase mouth. Tighten the bolts, finger-tight at first and then evenly and in a diagonal sequence until the recommended torque loading of 7.0 - 9.0 lbf ft (0.9 - 1.2 kgf m) is reached. Finally, temporarily plug the spark plug hole with a wad of clean rag to prevent the ingress of any contamination into the combustion chamber. The spark plug should not be fitted until the engine unit is refitted into the frame.

32.2a Fit and lubricate the small-end bearing

32.2b The 'Ex' mark must face the front of the engine

32.3 The circlip gap must be away from the piston cutouts

32.4 Ease the piston into the cylinder barrel

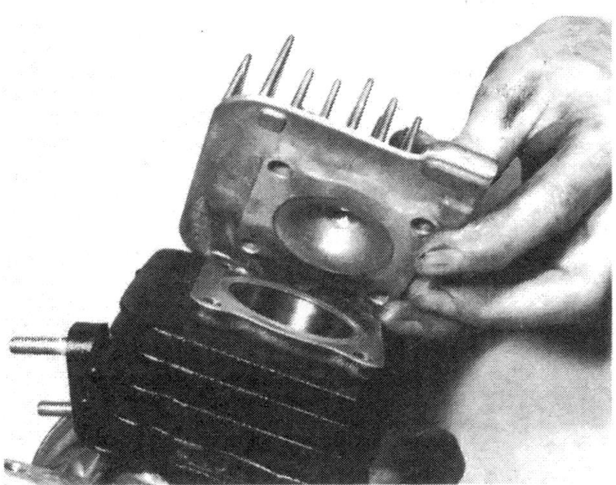
32.5 Fit the cylinder head with its new gasket

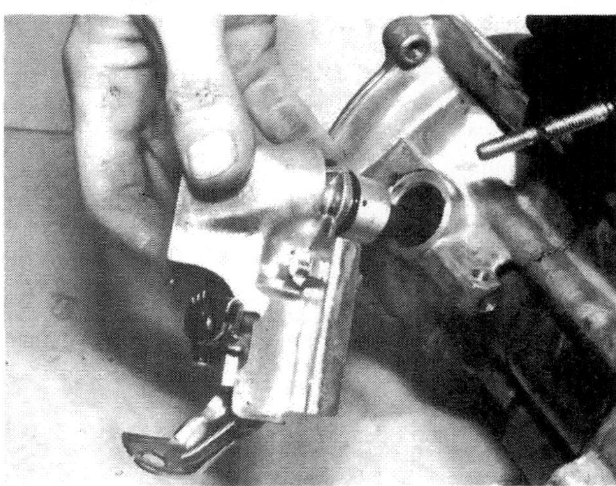
33.1 Insert the oil pump into its crankcase location

33 Reassembling the engine/transmission unit: fitting the oil pump

1 Before pushing the oil pump into its crankcase location, lightly lubricate with a lithium based grease the pump gear and the O-ring fitted to the pump boss. Insert the pump into the crankcase and carefully push it fully home. Fit and tighten the single pump retaining bolt.

34 Reassembling the engine/transmission unit: fitting the reduction gearbox components

1 The component parts of the reduction gearbox will have already been fitted to the gearbox cover (see Section 29 of this Chapter). This assembly should now be placed on a clean area of work surface adjacent to the gearbox casing and its components parts thoroughly lubricated with clean gearbox oil. Lubricate also the bearing and gearbox shaft locations within the gearbox casing.
2 Push the two gearbox cover locating dowels into their locations in the casing mating surface and position a new gasket in position over them. The cover assembly should now be fitted into the casing as far as is possible with hand pressure. The final fitting of the gearbox output shaft (rear wheel stub-axle) into its bearing will have to be accomplished by tapping the rear section of the cover with a soft-faced hammer until the shaft is driven fully home. Care must be taken whilst doing this to ensure that the thrust washer fitted to the intermediate shaft is not displaced and that the input and intermediate shafts enter their locations in the gearbox casing cleanly.
3 With the gearbox cover seated against the mating surface of the casing and the end tabs of the sealing gasket located correctly between the component mating surfaces, fit the four cover retaining bolts along with the oil drain bolt and its new sealing washer. Take care to tighten these bolts evenly whilst working in a diagonal sequence; this will greatly lessen the risk of the cover becoming distorted. Finally, check to ensure that the input and output shafts are free to rotate. Any binding of these shafts will necessitate separation of the gearbox casing so that the cause may be found and rectified.

34.2 Fit the gearbox cover assembly into the housing

34.3 The gasket sealing tabs must be correctly located

35 Reassembling the engine/transmission unit: fitting the flywheel generator

1 If the generator stator assembly was removed for the purposes of reconditioning the unit or to allow separation of the crankcase halves, it should now be refitted. Carefully thread the stator wires through the hole provided in the stator housing and push their retaining grommet into position in the hole. Position the stator plate in its housing and retain it in position by fitting and tightening its two retaining bolts.
2 Degrease the taper and thread of the right-hand crankshaft and also the bore of the flywheel generator rotor. Fit the Woodruff key into the keyway cut in the crankshaft taper and push the rotor into position over it. Fit the rotor retaining nut and tighten it, finger-tight.
3 It is now necessary to devise a means by which the rotor may be prevented from turning whilst its retaining nut is fully tightened to the recommended torque loading of 25 - 29 lbf ft (3.5 - 4.0 kgf m). Unfortunately, the method used for restraining the rotor during removal cannot be used if a socket is to be fitted over the retaining nut to enable a torque wrench to be used. It is, therefore, necessary to construct a simple tool from two lengths of mild steel bar and three nuts and bolts. The design of this tool and the method of its use is clearly illustrated in the figure accompanying this text.

36 Reassembling the engine/transmission unit: fitting the kickstart components and starter motor

1 Place the component parts of the kickstart assembly on a clean area of work surface adjacent to the transmission casing. Check that each part is clean and then smear it very lightly with clean gearbox oil. Do the same with the shaft locations within the transmission casing. The reason for oiling these parts is to protect them against the corrosive effect of any moisture build up in the transmission casing brought about by condensation. It must be realised that any excessive application of oil may lead to the drive assembly becoming contaminated.
2 Push the kickstart return stop into position over its locating stub and fit its retaining plate with the side marked 'Outside' facing upppermost. Fit and tighten the plate retaining bolt.
3 Apply a small amount of high melting point lithium based grease to the spline and spring groove of the idle pinion. Assemble the pinion, together with the two springs and the small locating pin onto the shaft and insert the complete assembly into the transmission casing whilst aligning the locating pin with the slot in the shaft location and the end of the friction spring with the guide channel in the transmission casing.
4 Fit the spring retaining pin into the end of the kickstart lever shaft. Push the bush into its location on the transmission casing and fit the shaft so that the punch mark on the toothed edge of its quadrant is in alignment with the corresponding mark on the boss of the idle pinion. Fit the kickstart return spring over the shaft and hook the end of the spring over the retaining pin. Grip the leg of the spring with a pair of pliers and locate its end beneath the retaining plate as shown in the accompanying photograph.
5 Degrease the surface of the left-hand crankshaft and the bore of the starter pinion. Push the pinion onto the shaft whilst pulling the outer end of the idle pinion shaft away from the pinion in order to prevent the teeth of the pinion from becoming caught on the edge of the shaft cutout. Place the end of a suitable length of tube against the boss of the starter pinion and using the tube as a drift in conjunction with a hammer, tap the pinion fully home onto the crankshaft.
6 Where a starter motor is fitted to the machine, check that the O-ring fitted to the starter motor boss is both serviceable and correctly located in its groove before pushing the motor into its crankcase location. It is considered good practice to renew this O-ring as a matter of course during a complete engine rebuild. If the O-ring is dry, lubricate it with a little engine oil before fitting the motor. Align the motor with the transmission casing and fit and tighten its two securing bolts.

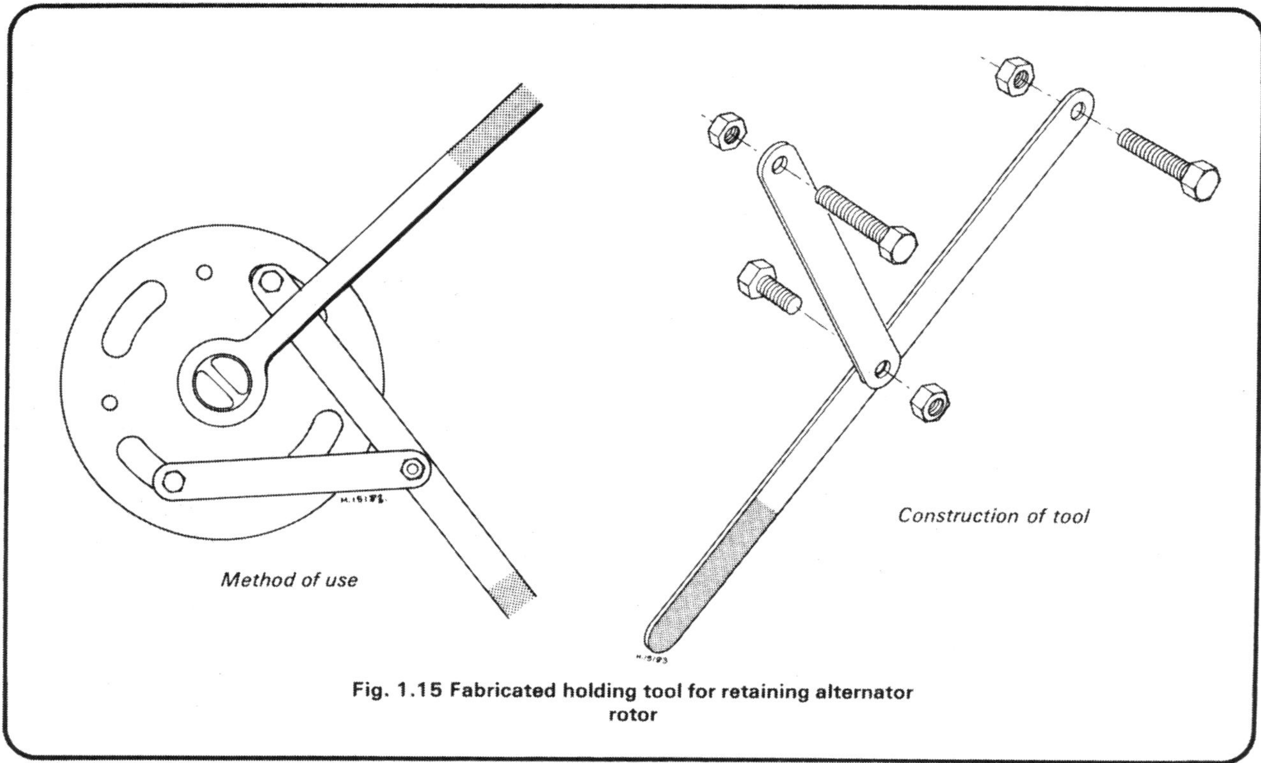

Fig. 1.15 Fabricated holding tool for retaining alternator rotor

35.1 Refit the stator plate into its housing

35.2a Refit the Woodruff key ...

35.2b ... and slide the flywheel rotor into position

36.2a Push the kickstart return stop into position ...

36.2b ... and fit its retaining plate

36.3a Align the kickstart idle pinion locating pin ...

36.3b ... and the end of the friction spring

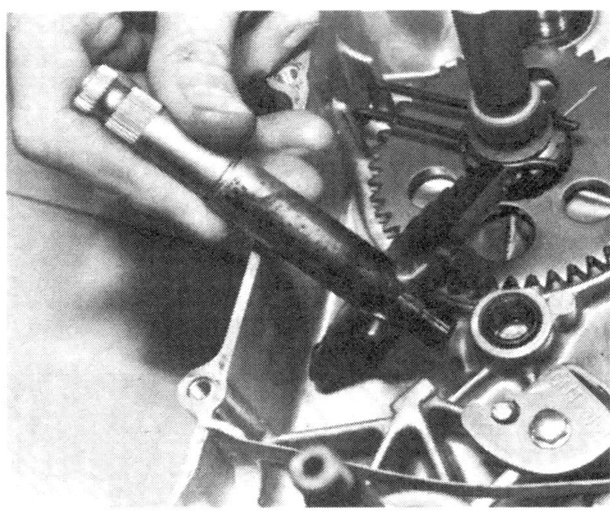
36.4a Fit the kickstart lever shaft ...

36.4b ... and align the two punch marks

36.4c Hook the return spring over the shaft pin ...

36.4d ... and locate its other end beneath the stop plate

36.5 Refit the starter pinion

37 Reassembling the engine/transmission unit: fitting the centrifugal clutch and crankshaft pulley

1 Degrease the surface of the gearbox input shaft and the bore of the clutch unit and push the unit down over the shaft as far as it will go with hand pressure. Place the end of a suitable length of tube against the clutch backplate retaining nut and using the tube as a drift in conjunction with a hammer, tap the clutch unit fully home onto the input shaft.
2 Position the clutch drum over the splines of the input shaft and fit and tighten finger-tight its retaining nut. It is now necessary to devise a means by which the clutch drum may be prevented from rotating whilst its retaining nut is fully tightened to the recommended torque loading of 25 – 29 lbf ft (3.5 – 4.0 kgf m). Unfortunately, the method used for restraining the drum during removal cannot be used if a socket is to be fitted over the retaining nut to enable a torque wrench to be used. It is, therefore, necessary to construct a tool with which to prevent the drum from rotating, and the method of doing this and of using the tool is fully described in Section 35 of this chapter which deals with fitting of the flywheel generator rotor.
3 Degrease the left-hand crankshaft and the bores of both the fixed and sliding drive pulley faces. Push the sliding face assembly onto the crankshaft so that it abuts against the starter pinion.
4 Push each of the two transmission cover locating dowels into their locations in the mating surface of the casing and carefully place the new sealing gasket in position over them. Fit the single thrust washer over the end of the kickstart lever shaft.
5 It is now necessary to carefully prise apart the two faces of the driven pulley with the flat of a large screwdriver so that the drive belt may be fitted between them and over the boss of the drive pulley sliding face. Take care, whilst doing this, to avoid causing damage to the belt with the screwdriver. Place the fixed face of the drive pulley in position on the crankshaft and fit and tighten finger-tight its retaining nut, with washer.
6 Using the tool described in Section 35 of this Chapter, hold the flywheel generator rotor in position and tighten the drive pulley retaining nut to a torque loading of 25 – 29 lbf ft (3.5 – 4.0 kgf m). Place the transmission cover in position over its locating dowels and tap around its edge with a soft-faced hammer until it seats on the gasket. Fit the nine cover retaining bolts in their previously noted positions. Remember to refit the guide for the rear brake cable beneath the head of the lower, most forward of these bolts. Commence tightening the bolts. This should be done evenly and in a diagonal sequence to preclude the risk of the cover becoming distorted.
7 Finally, note the alignment mark punched in the end of the kickstart lever shaft and the corresponding mark punched in the boss of the kickstart lever. Align these marks and push the lever onto the shaft. Fit and tighten the pinch bolt to retain the lever in position.

37.1 Position the centrifugal clutch assembly ...

37.2 ... followed by the clutch drum

37.3 Fit the sliding pulley face over the crankshaft

37.4a Fit the new transmission cover gasket over its locating dowels ...

37.4b ... and place the thrust washer over the kickstart lever shaft

37.5a Fit the drive belt ...

37.5b ... followed by the fixed pulley face and the transmission cover

37.6 Secure the transmission cover in position ...

37.7 ... and align the kickstart lever

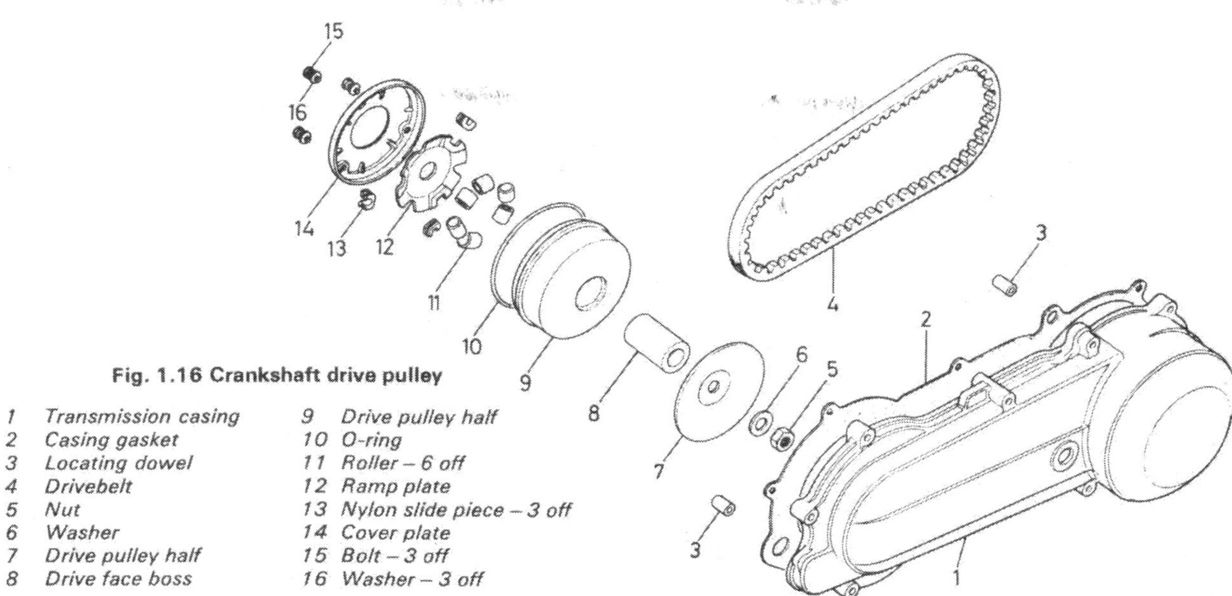

Fig. 1.16 Crankshaft drive pulley

1 *Transmission casing*
2 *Casing gasket*
3 *Locating dowel*
4 *Drivebelt*
5 *Nut*
6 *Washer*
7 *Drive pulley half*
8 *Drive face boss*
9 *Drive pulley half*
10 *O-ring*
11 *Roller – 6 off*
12 *Ramp plate*
13 *Nylon slide piece – 3 off*
14 *Cover plate*
15 *Bolt – 3 off*
16 *Washer – 3 off*

38 Reassembling the engine/transmission unit: fitting the reed valve assembly and carburettor

1 After having examined and, if necessary, renovated the reed valve assembly in accordance with the instructions given in Chapter 2, position the valve, sandwiched between its two new gaskets, on its crankcase location. Place the intake adaptor over the valve and retain it in position by fitting and tightening its four retaining bolts. Do this in a diagonal sequence whilst tightening in even increments to a torque loading of 6.0 - 9.0 lbf ft (0.8 - 1.2 kgf m). Note that the carburettor cover retaining bracket must be fitted beneath the head of the bolt shown in the photograph accompanying this text.

2 Check that the new O-ring is fitted properly into its retaining groove in the mating face of the carburettor and place the carburettor against the intake adaptor. Fit and tighten the two carburettor retaining nuts, with bolts, to a torque loading of 7.0 - 9.0 lbf ft (0.9 - 1.2 kgf m).

3 Refit the air filter housing over the carburettor mouth and secure it in position with its single retaining bolt.

38.1 Note the location of the carburettor cover retaining bracket

39 Reassembling the engine/transmission unit: fitting the fan assembly and cylinder head cowling

1 Place the cooling fan over the flywheel generator rotor and retain it in position with the four crosshead screws. Slide the left-hand section of cowling between the intake adaptor and cylinder barrel; positioning it so that the bolt holes in the base of the cowling align with those in the crankcase.

2 Position the choke control valve on the cylinder head and fit and tighten its two retaining screws. Reconnect each one of the pipes attached to the valve to its retaining stub on the intake adaptor or carburettor body and secure it in position with its spring clip. Remember that each pipe location is numbered to avoid confusion.

3 Fit the right-hand section of cowling over the cylinder head and position it so that its bolt holes align with those of the left-hand cowling half. Secure both cowling halves in position by fitting and tightening the three retaining bolts. Fitting of the trunking to the cylinder head cowling should be delayed until after a check for ignition timing has been carried out with the enigne fitted in the frame.

39.2 Secure each pipe end with its spring clip

40 Reassembling the engine/transmission unit: fitting the unit mounting plate

1 Full details of examining and renovating the unit mounting plate, as well as renewing the bonded-rubber bushes contained within the crankcase lugs, are contained within Chapter 4 of this Manual.
2 Slot the centre mounting plate into the crankcase and push the serviceable rubber buffer into position over the centre mounting stud. Place the unit mounting plate over the centre mounting stud and fit the second rubber buffer followed by the plate washer and retaining nut. Do not tighten this nut.
3 Align the unit mounting plate with the centre of each crankcase mounting and carefully drift the rear mounting bolt into position. Applying a light coating of grease to the shank of this bolt will ensure that it does not become corroded to the centre of the bonded rubber bushes, thus making extraction of the bolt at a later stage relatively trouble free.
4 Tighten the centre mounting nut to a torque loading of 14 - 22 lbf ft (2.0 - 3.0 kgf m). Fit the nut and washer to the rear mounting bolt and tighten the nut to a torque loading of 25 - 33 lbf ft (3.5 - 4.5 kgf m).

41 Reassembling the engine/transmission unit: fitting the rear wheel

1 Using a high melting point, lithium based grease, lubricate both the splines of the rear wheel stub-axle and the surfaces of the thrust collar. Push the collar into the recess of the transmission housing and slide the wheel into position over the stub-axle. Clean any excess grease off the threads of the stub-axle and fit the wheel retaining nut, together with its washer. With an assistant holding the wheel steady, tighten the retaining nut to a torque loading of 58 - 72 lbf ft (8.0 - 10.0 kgf m). Check that the wheel is free to rotate.

41.1 Do not omit to refit the thrust collar

42 Reassembling the engine/transmission unit: fitting the exhaust system

1 Commence fitting the exhaust system by pressing a new sealing ring into the recess of the exhaust port. Place the flange of the exhaust pipe over the two cylinder barrel studs and secure the silencer to the crankcase mounting lug by fitting and tightening finger-tight the single bolt, with its plain washer.
2 Tighten the two flange nuts which retain the exhaust pipe to the cylinder barrel and then tighten the silencer retaining bolt. Fitting of both the exhaust pipe and silencer guards will have to be delayed until after the ignition timing has been checked and the trunking to the cylinder head cowling refitted.

42.1 Press a new sealing ring into the exhaust port recess

43 Refitting the engine/transmission unit into the frame

1 It is worth checking at this stage that nothing has been omitted during the reassembly sequences. It is better to discover any left-over components at this stage rather than just before the engine is due to be started.
2 Refitting of the engine/transmission unit into the frame is, generally speaking, a direct reversal of the removal sequence. As with removal, it may be found advantageous to have an assistant present to steady the machine whilst the engine unit is eased into position.
3 Commence refitting the engine unit by selecting the unit pivot bolt and applying a light coating of a high melting point, lithium based grease to its shank. Doing this will ensure that the bolt does not become corroded to the centre of the bonded rubber bushes in the unit mounting plate, thus making extraction of the bolt at a later stage relatively trouble free.
4 Carefully ease the engine unit into position in the frame so that the unit mounting plate is aligned correctly between the two frame plates. Working from the left-hand side of the machine, push the pivot bolt through the mounting plates and tap it fully home with a soft-faced hammer. Fit the nut and washer to the bolt and tighten the nut to a torque loading of 25 - 33 lbf ft (3.5 - 4.5 kgf m).
5 Refit the rear suspension unit to the transmission casing and tighten its retaining bolt to a torque loading of 18 - 25 lbf ft (2.5 - 3.5 kgf m). Check tighten the nut which retains the suspension unit to the frame to a torque loading of 22 - 29 lbf ft (3.0 - 4.0 kgf m).
6 Re-route the rear brake operating cable along the underside of the transmission casing. The cable must be located correctly in both the guide clamp at the front of the casing and the anchor plate at the rear of the casing. Make sure that both the clamp and plate retaining bolts are fully tightened. With the cable routed through the trunnion of the brake cam operating lever, fit the adjuster nut to the cable end. This nut should be tightened until the amount of free play measured at the tip of the handlebar lever is 10 - 15 mm (0.4 - 0.6 in).
7 Remove the wad of rag from the spark plug hole in the cylinder head and fit and tighten the spark plug. Reconnect the HT lead suppressor cap to the spark plug. Reconnect all disturbed electrical connections whilst taking note of their colour coding.

8 Lightly lubricate the surface of the carburettor valve with clean engine oil and carefully relocate the valve assembly in the carburettor body. Ensure that the jet needle does not become caught on the edge of the needle jet as the valve assembly is pushed into position and check that the groove in the valve is in alignment with the tip of the throttle stop screw. Tighten the top of the mixture chamber.

9 With the carburettor valve assembly refitted, check that the throttle cable is correctly adjusted. This may be done by measuring the amount of free play at the throttle twistgrip flange. If the measurement obtained is outside the limits of 2 - 6 mm (0.08 - 0.24 in), then the cable must be adjusted by turning the adjuster at the handlebar end of the cable. If the correct amount of free play cannot be obtained by this method of adjustment, then it is recommended that the cable be renewed. Do not omit to retighten the adjuster locknut on completion of adjustment.

10 Relocate the oil pump control cable adjuster in its retaining bracket and fit a serviceable rubber gaiter over the cable end. Push the oil pump lever towards the cable and reconnect the cable inner to the lever. The control cable should now be checked for adjustment by opening the throttle as far as possible and then checking that the pointer on the pump lever is in exact alignment with the corresponding pointer cast in the pump body. Note that this check should only be carried out with the throttle cable correctly adjusted. If adjustment is found to be incorrect, move the adjuster body through its retaining bracket until both pointers come into alignment. This can be done by rotating the adjuster retaining nuts in the required direction. With the pointers aligned, secure the adjuster in its retaining bracket by tightening the two nuts against each other.

11 It is now necessary to bleed the oil pump of air. In order to do this, it is necessary to have either a syringe or an oil container with a long spout fitted to it, either one of which should be filled with clean engine oil. Check that the oil tank is full. Place the end of the oil feed pipe over a piece of absorbent rag and remove the plug from its end. Allow oil to drain from the pipe until it is seen to be free of air. Refit the plug to the pipe. Take the oil container and press its spout against the inlet stub of the oil pump. Pump as much oil into the pump as possible, unplug the end of the feed pipe and quickly push it over the inlet stub. Retain the pipe in position with its spring clip.

12 With the oil pump and oil feed pipe thus primed with oil, it is now necessary to prime the oil outlet pipe which runs from the pump to the intake adaptor. Push the end of this pipe over the stub of the intake adaptor and secure it in position with its spring clip. Raise the free end of the pipe above the level of its fixed end and, using the oil container, fill the pipe with clean engine oil. Remove the container from the pipe end and pinch the end of the pipe together between two fingers. Reconnect the pipe to the outlet stub of the oil pump whilst taking care not to allow any loss of oil from the pipe. Secure the pipe to the stub with its spring clip. No more can be done towards bleeding of the oil pump until the engine is ready to be started.

13 Reconnect the fuel feed pipe to the horizontal stub of the vacuum-operated fuel tap and the vacuum pipe to the vertical stub of the tap. Secure each pipe in position with its spring clip.

14 Refit both the rear mudguard and the centre body panel. Replace the battery in its tray whilst making sure that its vent tube is both secure and correctly routed. Reconnect the battery leads whilst observing the polarity markings. Ensure that both battery terminals are free from corrosion and protect them with a liberal coating of petroleum jelly. Do not use ordinary grease. Refit the battery cover and securing strap.

15 Remove the oil level bolt, with its sealing washer, from the cover of the reduction gearbox. Check the condition of the sealing washer and renew it if necessary. Refill the gearbox with SAE 10W/30 oil by pouring the oil in through the level hole. Approximately 90 cc (3.16 Imp fl oz) of oil should be needed. The oil level is correct when oil is seen to exude back out of the oil level hole. Refit and tighten the oil level bolt with its sealing washer.

16 Finally, carry out a thorough check around the engine and its associated frame components to ensure that all disturbed components have been fitted correctly and are functioning properly. No control cable, electrical lead, etc should be allowed to be trapped between any moving components or be in contact with hot engine castings.

43.11 The oil tank must be refilled

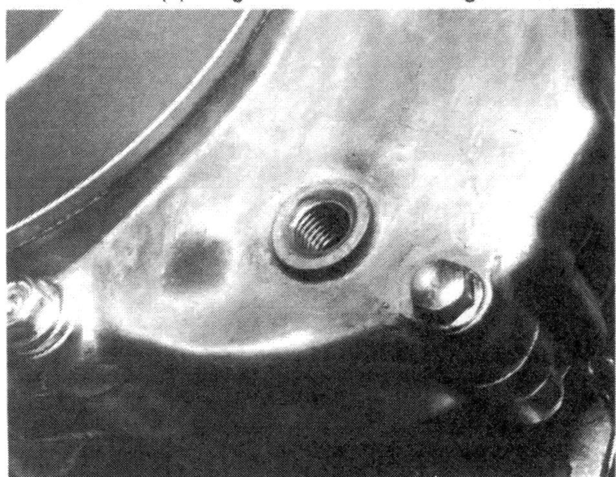
43.15 Remove the gearbox oil level bolt

43.16 Check that all disturbed components have been fitted correctly

44 Starting and running the rebuilt engine

1 When the initial start-up is made, run the engine slowly for the first few minutes, especially if the engine has been rebored or a new crankshaft fitted. Check that all the controls function correctly and that there are no oil or fuel leaks. The exhaust will emit a high proportion of white smoke during its initial running period, as the excess oil used whilst the engine was re-assembled is burnt away. The volume of smoke should gradually diminish until only the customary light blue haze is observed during normal running.

2 Special attention should be paid to the presence of air in the outlet pipe from the oil pump directly the engine is started. Because of the method used to prime the outlet pipe, it is impossible to avoid a small amount of air getting into the pipe. In practice, this air was seen to take the form of a bubble which was approximately one tenth of an inch in length. This bubble gradually moved towards the intake adaptor during the pump bleeding procedure and was eventually expelled from the pipe to be followed by a constant flow of air-free oil from the pump. The presence of this small amount of air in the outlet pipe was considered to be acceptable.

3 Complete the oil pump bleeding procedure by holding the lever of the pump in its fully-open position with the tip of one finger whilst observing the oil outlet pipe at its union with the pump for any sign of air issueing from the pump. The presence of a constant flow of air from the pump will indicate the need for the pump bleeding procedure to be repeated. Allow the pump lever to return to its normal position directly all air has been expelled from the outlet pipe. Increasing the throttle opening for short intervals will help to move any air out of the outlet pipe but care must be taken not to race the engine unnecessarily.

4 As the ignition timing of the Honda Melody is fixed, there should be no need to check the timing unless the engine is thought to be suffering from a fault in the ignition system. Full details of checking the ignition timing and carrying out tests on the ignition system components may be found in Chapter 3 of this Manual. On completion of any checks, refit the trunking to the cylinder head cowling, the plastic grid to the trunking, the heat guard to the exhaust pipe and the heat guard to the exhaust silencer.

5 Unless the carburettor was dismantled for examination and renovation during the engine servicing procedure, there should be no need to carry out any alteration of its settings; in which case the carburettor and air filter assembly shield may be secured in position by fitting and tightening its single retaining screw. Full details of resetting the carburettor are given in Chapter 2 of this Manual.

6 Note that it is wise to carry a spare spark plug of the correct type during the first run of the machine on the road, as the existing plug may oil up due to the temporary excess of oil in the combustion chamber. Before riding the machine away, check to ensure that the rear brake is functioning correctly.

7 Remember that a good seal between the piston and the cylinder barrel is essential for the correct function of the engine. A rebored two-stroke engine will require more careful running-in, over a longer period, than its four-stroke counterpart. There is a far greater risk of engine seizure during the first hundred miles if the engine is permitted to work hard.

8 Do not tamper with the exhaust system or use a holed or damaged silencer. Unwarranted changes in the exhaust system will have a very marked effect on engine performance, invariably for the worse. The same advice applies to dispensing with the air cleaner.

9 Do not, on any account, add extra oil to the fuel in the mistaken belief that a little extra oil will improve the engine lubrication. Apart from creating excess smoke, the addition of oil will make the mixture much weaker, with the consequent risk of overheating and engine seizure.

10 If at any time a lubrication failure is suspected, stop the engine immediately and investigate the cause. If an engine is run without oil, even for a short period, irreparable engine damage is inevitable.

11 On completion of the initial run, allow the engine to cool down completely and then carry out a final check for security of components, fuel and oil leaks, etc.

44.4 Note the air trunking retaining clip

45 Fault diagnosis: engine

Symptom	Cause	Remedy
Engine will not start	Defective spark plug	Remove plug and lay it on cylinder head Check whether spark occurs when engine is kicked over
	Defective ignition system component	Carry out checks described in Chapter 3
	Fuel tank empty	Refill
Engine runs unevenly	Ignition and/or fuel system fault	Check as though engine will not start
	Blowing cylinder head joint	Oil leak should provide evidence. Check for warpage
	Choked silencer	Remove and clean
Lack of power	Fault in fuel system	Check system
	Choked silencer	See above
White smoke from exhaust	Engine needs rebore	Rebore and fit oversize piston
	Tank contains two-stroke petroil	Drain and refill with straight petrol
	Too much oil entering engine	Check oil pump setting
Engine overheats	Pre-ignition and/or weak mixture	Check carburettor settings also grade of plug fitted
	Lubrication failure	Stop engine. Check level of engine oil in tank Check oil pump setting

46 Fault diagnosis: transmission

Symptom	Cause	Remedy
No drive between engine and rear wheel	Slipping or broken drive belt	Renew belt
	Reduction gearbox pinions disengaged or gearbox broken	Dismantle and overhaul
Engine runs normally but road speed abnormally slow	Drive belt worn	Renew belt
	Centrifugal clutch shoe linings worn	Dismantle and overhaul clutch unit
	Variable ratio pulley assembly jammed	Dismantle and overhaul drive and driven pulley assemblies
Machine creeps forward with engine running at idle speed	Centrigual clutch shoe return springs weakened or stretched	Dismantle clutch unit and renew springs

Chapter 2 Fuel system and lubrication

Refer to Chapter 7 for information on the NB, ND and NP50 models

Contents

Specifications

Fuel tank

Overall capacity	3.2 litre (0.71 Imp gal)
Reserve capacity	0.5 litre (0.88 Imp pint)

Carburettor

Make	Keihin
ID number	PD19A-A
Venturi diameter	12 mm (0.47 in)
Float height	12.2 ± 1.0 mm (0.48 ± 0.04 in)
Pilot air screw setting	$1\frac{3}{4}$ turns out from fully in
Engine idle speed	1800 ± 150 rpm
Throttle twistgrip free play	2 – 6 mm (0.07 – 0.23 in)
Needle clip position	2nd groove from top

Lubrication system

Type	Pump fed from separate oil tank
Oil type	Honda 2-stroke injector oil or equivalent
Tank capacity	1.2 litre (2.1 Imp pint)

1 General description

The fuel system fitted to the Honda Melody is unusually sophisticated for a small capacity two-stroke machine, featuring a number of rather complex devices which serve to remove a number of minor manual controls. The first of these devices is an automatically operated choke control valve which obviates the need for the normal hand operated, cable/lever devices found on their small Honda machines. The second device is a vacuum operated fuel tap which is brought into operation directly the engine is turned by operating the starter motor (where fitted) or the kickstart. This tap replaces the usual hand operated type and therefore obviates the need to remember to turn the tap on or off. The carburettor is of the normal slide type but features the adaptors and drillings necessary for operation of the choke control valve.

The fuel/air mixture is admitted to the engine via a reed valve assembly which allows more accurate timing of the mixture than conventional piston porting alone. The valve is entirely automatic in operation, opening and closing according to changes in pressure inside the crankcase.

Lubrication of the engine is by pump, a precisely metered amount of two-stroke oil being injected via a small nozzle in the intake adaptor. The oil feed rate is varied according to engine speed and also by the throttle opening, the pump output being controlled by a cable which is interconnected to the throttle cable.

2 Fuel tank: removal and refitting

1 Gain access to the tank mounting points by removing the centre body panel from the machine. This panel is retained in position by a single dome head nut with washer and a single screw. Detach the seat from the tank by removing the two nuts which secure its hinge to the tank top.
2 It is now necessary to disconnect the fuel and vacuum pipes from the fuel tap. Before doing this, take note of the risks involved and note the following safety precautions. Take great care to place the machine in a well ventilated space where there is no chance of a build-up of fuel vapour occurring with the subsequent risk of a fire or explosion. Drain any fuel from disconnected pipes into a clean metal container that is equipped with a properly sealed lid; never use an open or plastic container. Store both this container and the removed tank in a well ventilated open space. Never smoke or expose the machine to any kind of naked flame whilst removing or fitting the fuel tank.
3 Detach the fuel and vacuum pipes from the fuel tap by releasing their spring clips and pulling them off their retaining stubs. Note that the pipe attached to the horizontal stub of the tap is the one that supplies fuel to the carburettor. Drain any fuel retained in this pipe into a container. If fuel continues to issue from the fuel tap, then the tap is defective and must be serviced as described in Section 4 of this Chapter.
4 Trace the electrical wires from the float switch fitted to the top of the tank to their nearest push connectors. Unplug these connectors and place them on the tank top. With the tank properly supported, remove each of its four mounting bolts and then lift the tank clear of the frame.
5 Fitting the tank is a direct reversal of the removal procedure. Inspect the fuel and vacuum pipes for deterioration or damage before reconnecting them and renew each one as necessary. Note that the electrical wires to the float switch are colour coded to avoid confusion whilst reconnecting. Once the tank is fitted in the machine, start the engine and allow it to run at tick-over speed whilst carrying out a thorough check around the pipe connection points for any signs of fuel leakage. If a leak is found, it must be cured before the machine is ridden, otherwise there is a considerable risk of fire resulting in serious injury to the rider.

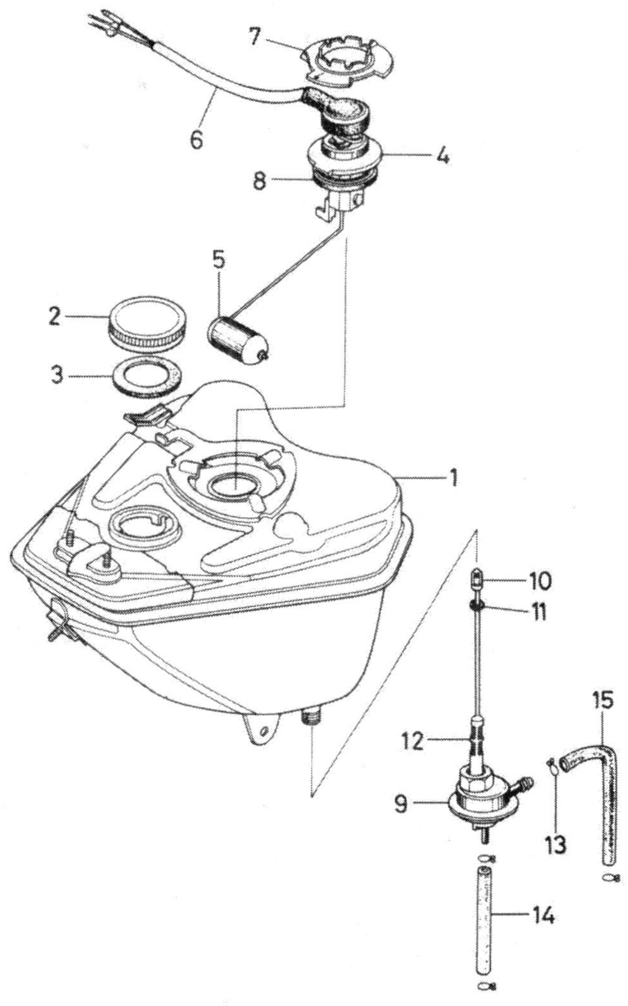

Fig. 2.1 Fuel tank

1 Fuel tank
2 Filler cap
3 Gasket
4 Float switch
5 Float
6 Float switch wiring
7 Retaining ring
8 Seal
9 Fuel tap
10 Air filter
11 Circlip
12 Filter
13 Pipe clip – 2 off
14 Vacuum pipe
15 Fuel delivery pipe

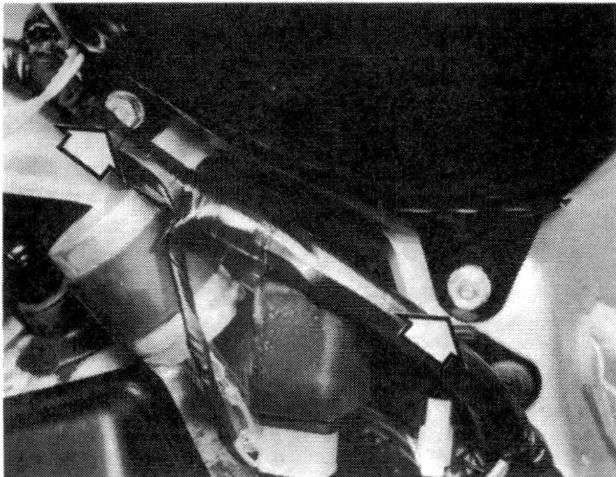

2.1 Gain access to the fuel tank mounting points (arrowed)

3 Fuel tank: flushing

1 The fuel tank may need flushing out occasionally to remove any accumulated debris which inevitably builds up over the years. This is especially true if water has contaminated the fuel, as this can cause persistent and annoying running problems as it gets drawn into the carburettor.
2 Flushing is best done by first removing the tank, as detailed in the preceding Section, and then draining the tank of any contaminated fuel. Any debris or water may now be cleaned out by flushing the tank with clean fuel. Note that this operation must be undertaken outdoors and away from naked flames or lights, otherwise serious personal injury may result from the ignition of the fuel vapour.

4 Fuel tap: fault diagnosis and rectification, removal and refitting

1 The fuel tap may be checked for correct operation by carrying out the following tests. Gain access to the tap by removing the centre body panel from the machine. This panel is retained in position by a single dome head nut with washer and a single screw. Before disconnecting any part of the tap assembly, take note of the following safety precautions. Take great care to place the machine in a well ventilated space where there is no chance of a build-up of fuel vapour occurring with the subsequent risk of a fire or explosion. Drain any fuel into a clean metal container that is equipped with a properly sealed lid; never use an open or plastic container. Store this container in a well ventilated open space. Never smoke or expose the machine to any kind of naked flame whilst draining fuel.
2 If difficulty is experienced in starting the engine and fuel starvation is suspected, then detach the vacuum pipe from the intake adaptor and the fuel feed pipe from the carburettor, having first removed the carburettor shield to gain access. Have a container nearby in which to catch any fuel that may drain from the fuel pipe.
3 Test the operation of the fuel tap by placing the end of the fuel pipe in the container and then applying suction to the end of the vacuum pipe to cause the diaphragm in the tap to move downwards against its return spring. Once this diaphragm is moved, fuel will flow from the tank, through the tap and out of the end of the fuel pipe. If this does not happen, then the fuel starvation can be attributed to one of the following causes:

A stuck fuel tap diaphragm
A blocked vacuum pipe
A blocked fuel pipe
A blocked fuel tap filter

It should be noted that only a gentle amount of suction need be applied to the end of the vacuum pipe to cause operation of the tap diaphragm. This suction may be applied by placing the pipe in one's mouth and sucking. Some care should be taken whilst doing this to ensure that there is no fuel present in the vacuum pipe as would be the case if the tap diaphragm were split. An alternative method of applying suction would be to use a foot or hand pump.
4 If operation of the tap is found to be at fault, then detach both the fuel and vacuum pipes from the tap and check each one for any sign of a blockage in its bore. Clear any blockage found by blowing through the pipe with compressed air. Reconnect the pipes to the fuel tap and secure them with their spring clips.
5 It is possible to free a tap diaphragm that is stuck by applying a jet of compressed air to the end of the fuel pipe. This will cause the diaphragm to be pushed down against its spring, thereby freeing it from the closed position. Take care not to use too great a pressure of air when doing this otherwise the diaphragm will be permanently damaged. If the diaphragm will not free or if fuel is seen to issue from the end of the vacuum pipe, then the tap is defective and should be renewed.
6 In order to gain access to the fuel tap filter for the purposes of inspection and cleaning, it is first necessary to completely drain the tank of fuel and then to remove the fuel tap from the tank. Draining of the tank can be achieved by either applying suction to the vacuum pipe in order to allow fuel to pass from the tank into a container or by removing the tank and inverting it to allow fuel to drain out of the fuel filler point.
7 With the tank thus drained, unscrew the locknut which serves to retain the fuel tap in position and carefully pull the tap out of its location whilst taking care not to cause damage to the filter stack. Draw the small air filter off the end of the fuel tap air pipe and carefully remove the clip located beneath it. The filter can now be removed from the tap together with its base gasket.
8 Clean the fuel filter by rinsing it in clean fuel. Any stubborn traces of contamination can be removed by gently brushing the filter with a soft-bristled brush which has been soaked in fuel; a used toothbrush is ideal. Complete the cleaning process by blowing the filter clear with a jet of compressed air. Remember to take the necessary fire precautions whilst carrying out this cleaning procedure and always wear eye protection against any fuel that may spray back from the brush or air jet. On completion of cleaning, closely inspect the filter for any splits or holes that will allow the passage of sediment through it and into the carburettor. Renew the filter if it is in any way defective.
9 Inspect and, if necessary, renew the small base gasket to the filter. Carefully clean the small air filter with a jet of compressed air and carry out a general inspection of the fuel tap for any signs of damage. Make sure that the air pipe is clear of contamination and fit the fuel filter with its base gasket over it. Push the fuel filter down into the centre of the tap locknut to seat it and then fit to the air pipe the support clip for the air filter. This clip must be located so that it is 17 mm (0.67 in) from the top of the air pipe. Fit the air filter and then carefully relocate the fuel tap in the tank. Hold the tap so that its fuel and vacuum pipe retaining stubs are correctly positioned and then tighten the locknut. Pour a small amount of fuel into the tank and check for any leakage of fuel from around the tap to tank joint. if fuel is seen to leak from this joint and the leak cannot be stopped by further tightening of the nut, then it will be necessary to remove the tap and relocate or renew the fuel filter base gasket. Do not overtighten the tap locknut.
10 Finally, after having fitted the fuel tap and reconnected the fuel and vacuum pipes, refill the fuel tank and start the engine. Allow the engine to run at tick-over speed whilst carrying out a thorough check for leaks at all of the disturbed connection points. If a leak is found, it must be cured before the machine is ridden, otherwise there is a considerable risk of fire resulting in serious injury to the rider. On completion of a satisfactory leak check, refit the carburettor shield and the centre body panel.

4.1 The vacuum operated fuel tap

5 Fuel and vacuum pipes: examination and renovation

1 The condition of the fuel and vacuum pipes connected to the fuel tap should be checked periodically. The synthetic rubber pipes are quite resistant to deterioration, but may eventually develop leaks where they push over their respective stubs. This can be corrected by disconnecting the pipe and slicing off the worn end. For obvious reasons this cannot be repeated many times before renewal becomes necessary.

2 Always obtain replacement pipes of the correct type, especially where identification marks are present. Replacement pipes of different materials should be avoided, and natural rubber tubing must be avoided at all costs. This latter material is attacked by fuel and will disintegrate internally blocking the carburettor jets with a thick rubbery sludge.

6 Choke control valve: removal, examination, renovation and refitting

1 The choke control valve is fitted to provide the necessary fuel-rich mixture required for cold starting. This it does automatically by means of a spring-loaded plunger and a bi-metallic strip. With the engine cold, the plunger is set in the 'On' position to allow a fuel-rich mixture. As the engine reaches normal operating temperature, the heat sensing bi-metallic strip begins to move, drawing the plunger with it and thereby closing the cold start circuit.
2 Before suspecting the choke control valve of being defective, detach the carburettor shield and carry out a close examination of all the vacuum pipes leading to the unit. Leakage of any one of these pipes will result in malfunction of the valve which in turn will mean that once the engine reaches its normal operation temperature, it will begin to run unevenly due to the mixture being incorrect. Check the pipe connections first, paying particular attention to the area near the various mounting stubs. If deterioration of a pipe end is discovered, cut off the affected length of pipe before reconnecting the pipe to restore the seal. Check that each spring clip is still serviceable and has not become fatigued. Check also that the choke control filter is not clogged.
3 If it is decided that the choke control valve is at fault, then commence removal by first detaching the air filter housing from the carburettor. This housing is retained in position by a single bolt which passes through its lower edge, just to the rear of the carburettor float chamber.
4 It will be seen that the head of the innermost of the two bolts which retain the choke control valve in position is obscured by the right-hand half of the cylinder head cowling. It will, therefore, be necessary to remove this section of cowling to allow withdrawal of the bolt and removal of the valve. Commence the necessary dismantling procedure by detaching the centre body panel. This panel is held in position by one dome head nut, with washer, and a single screw.
5 Remove the exhaust system. Full information on doing this is contained in Section 16 of this Chapter. Detach the HT lead suppressor cap from the spark plug and remove the plug. Removal of the air trunking and cylinder head cowling half is simply a matter of working in a logical sequence, commencing with the two cross head screws which retain the plastic grid in position and then detaching the trunking which should be followed in turn by the cowling half.
6 Having thus fully exposed the choke control valve, unscrew its two retaining bolts and detach it from the cylinder head. Disconnect each of the three pipes attached to the valve from their retaining stubs on the carburettor body and intake adaptor. Do this by releasing each pipe retaining clip before pulling the pipe from position. It will be seen that each one of the three pipe retaining stubs on the valve body is numbered. These numbers correspond with identical numbers cast in the bodies of both the carburettor and intake adaptor. As long as each pipe is left connected to either one of its retaining stubs, there should be no confusion as to its fitted position when the time comes for reconnection of the valve.
7 Place the choke control valve on a clean area of work surface and remove the two small cross head screws which retain its cover in position. Carefully detach the cover from the body of the valve whilst taking care not to tear the seal placed between the two components. Closely inspect the bi-metallic strip for fatigue or failure. If this strip is found to be serviceable, then it may be removed by unscrewing its two retaining screws to permit examination of the spring and plunger assembly. If any one item in the valve is found to be defective, then the complete valve assembly must be renewed as Honda do not list the separate component parts for the valve. If in doubt as to the condition of the valve, then return it to a Honda service agent for further examination and, if necessary, renewal.
8 Fitting of the choke control valve is a direct reversal of the removal procedure. Remember that each pipe location is numbered and take care to ensure that each pipe is retained in position by its spring clip.

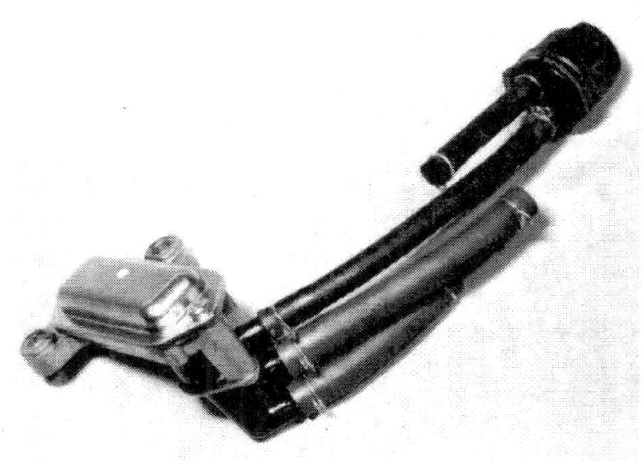
6.7a Place the choke control valve on a clean work surface ...

6.7b ... and closely inspect the bi-metallic strip

7 Carburettor: removal and refitting

1 Remove the carburettor shield by unscrewing the single retaining screw which passes through its forward edge. Unscrew the carburettor and carefully pull the valve assembly out of the carburettor. Secure this assembly, still connected to the cable, to a point on the machine which is clear of the carburettor body.
2 Unscrew the single bolt which serves to retain the air filter housing in position. This bolt is located beneath the housing, just to the rear of the carburettor float chamber. With the bolt removed, pull the filter housing back to clear the carburettor mouth before lifting it clear of the engine unit.

3 Detach each of the choke control pipes from their retaining stubs on the carburettor body. Obtain a small clean container and release the fuel feed pipe retaining clip from the pipe end attached to the carburettor. Pinch the end of the pipe together, pull it off its retaining stub and allow any fuel contained within the pipe to drain into the container. Take care to observe the necessary fire precautions whilst doing this. The carburettor can now be detached from the intake adaptor by removing its two retaining bolts.
4 Fitting the carburettor is a straightforward reversal of the removal procedure, whilst noting the following points. Before placing the carburettor in position against the intake adaptor, check that the O-ring fitted to its mating face is both serviceable and located correctly in its retaining groove. Fit and tighten the two carburettor retaining nuts, with bolts, to a torque loading of 7.0 – 9.0 lbf ft (0.9 – 1.2 kgf m).
5 Reconnect the fuel feed and choke control pipes to the carburettor. Retain each pipe in position with its spring clip and note that the retaining stubs for the choke control pipes are numbered to avoid confusion. Refit the air filter housing over the carburettor mouth and secure it in position with its single retaining bolt.
6 Lightly lubricate the surface of the carburettor valve with clean engine oil and carefully relocate the slide assembly in the carburettor body. Ensure that the jet needle does not become caught on the edge of the needle jet as the valve assembly is pushed into position and check that the groove in the slide is in alignment with the top of the throttle stop screw. Tighten the top of the mixture chamber.
7 With the carburettor valve assembly refitted, check that the throttle cable is correctly adjusted. This may be done by measuring the amount of free play at the throttle twistgrip flange. If the measurement obtained is outside the limits of 2 – 6 mm (0.08 – 0.24 in), then the cable must be adjusted by turning the adjuster at the handlebar end of the cable. If the correct amount of free play cannot be obtained by this method of adjustment, then it is recommended that the cable be renewed. Do not omit to retighten the adjuster locknut on completion of adjustment or to synchronize the oil pump control cable.
8 Before refitting the carburettor shield, start the machine and allow the engine to run at tick-over speed. Carry out a check for signs of fuel leakage, especially at the disturbed fuel feed pipe connection point. If a leak is found, it must be cured before the machine is ridden, otherwise there is a considerable risk of fire resulting in serious injury to the rider.

7.1 Remove the carburettor shield

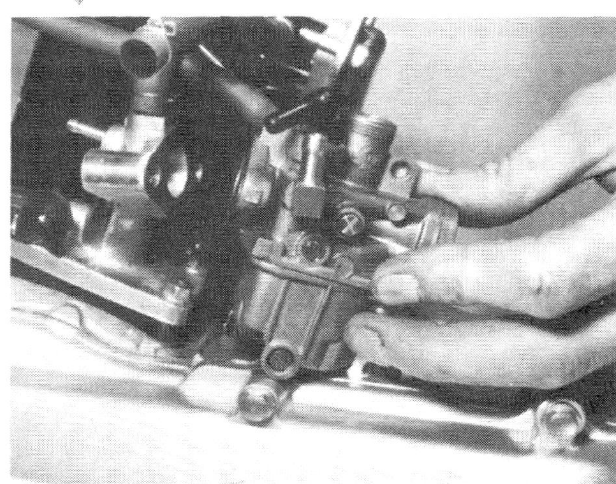
7.4 Position the carburettor against the intake adaptor

7.5 Reconnect the fuel feed and choke control pipes

7.6 Carefully relocate the carburettor slide

8 Carburettor: dismantling, examination, renovation and reassembly

1 Before dismantling the carburettor, cover an area of the work surface with clean paper or rag. This will not only prevent any components that are placed upon it from becoming contaminated with dirt, moisture or grit but, by making them more visible, will also prevent the many small components removed from the carburettor body from becoming lost.
2 Proceed to dismantle the carburettor by holding it over a small clean container and then removing the drain screw from the base of the float chamber. Any fuel contained in the float chamber should be allowed to drain into the container. Take care to observe the necessary fire precautions whilst doing this and during the various cleaning procedures listed in the following paragraphs of this Section.
3 Unscrew the two screws which retain the float chamber to the carburettor body. Note the condition of the spring washer fitted beneath the head of each of these bolts and renew each one if it is found to be flattened or broken. Carefully separate the float chamber from the carburettor body whilst taking great care not to stretch or tear the seal between the two mating surfaces.
4 Remove the domed cross head screw which serves to retain the float pivot pin in its groove in the carburettor body. Detach the float, with its pivot pin and needle, from the carburettor body. The float needle should now be displaced from the float and put aside in a safe place for examination at a later stage. The needle is very small and easily lost.
5 Locate the pilot air screw and screw it inwards until it is felt to come into contact with its seat. Count the number of turns required to do this and take care not to overtighten the screw. Make a note of the screw setting and then remove it. Carry out a similar procedure to remove the throttle stop screw. Failure to note the settings of the aforementioned screws will make it less easy to 'retune' the carburettor after it has been reassembled and refitted to the machine.
6 Carefully remove the O-ring from the flange of the carburettor body and the seal from its retaining groove in the float chamber. Closely inspect each item for signs of damage and deterioration and renew each one as necessary. Note that it is considered good practice to renew both of these seals as a matter of course when the carburettor is being renovated. Both the body of the carburettor and the float chamber are now devoid of all removable component parts, the jets of the carburettor being pressed into their locations and thus forming what must be considered to be a permanent part of the carburettor body. The vent tubes connected to the float chamber need not be removed unless they are seen to be perished or in any way damaged and therefore require renewal.
7 Prior to examination of the carburettor component parts, clean each part thoroughly in clean fuel before placing it on a piece of clean rag or paper. Use a soft nylon-bristled brush to remove any stubborn contamination from the castings and blow dry each part with a jet of compressed air. Avoid using a piece of rag for cleaning since there is always risk of particles of lint obstructing the airways or jet orifices. Never use a piece of wire or any pointed metal object to clear a blocked jet, it is only too easy to enlarge a jet under these circumstances and increase the rate of petrol consumption. If an air line is not available, a blast of air from a tyre pump will usually suffice. If all else fails to clear a blocked jet, remove a bristle from the soft-bristled brush and carefully pass it through the jet to clear the blockage.
8 Check each casting for cracks or damage and check that each mating surface is flat by laying a straight-edge along its length. A distorted casting must be replaced with a serviceable item.
9 Ensure that the O-ring fitted to the flange of the carburettor body and the seal fitted to the float chamber are both correctly located in their retaining grooves. Examine and, if necessary, renew the small O-ring fitted to the drain screw. The springs fitted to the throttle stop and pilot air screws should now be carefully inspected for signs of fatigue and corrosion and renewed if necessary.
10 The seating area of the float needle will wear after lengthy service and should be closely examined with a magnifying glass. Wear usually takes the form of a ridge or groove which will cause the float needle to seat imperfectly. If the needle has to be renewed, remember that the needle seat will have worn in unison and in extreme cases, will also need to be renewed. Note that the needle seat forms part of the carburettor body which means that if the seat is defective then the complete carburettor body will have to be renewed. Check also that the small pin which protrudes from the end of the needle is free to move and is returned to its extended position by the action of the spring fitted beneath it. The correct action of this pin is essential in cushioning the movement of the float against the needle which in turn acts to reduce the amount of wear on the seating area of the needle.
11 Closely examine the float for signs of damage or leakage. Leakage of the float will be obvious on inspection, because the float material is translucent. A defective float must be renewed.
12 Move to the machine and inspect the throttle valve for wear. This wear will be denoted by polished areas on the external diameter. Excessive wear will allow air to leak past the valve, thereby weakening the fuel/air mixture and producing erratic slow running. Many mysterious carburation maladies may be attributed to this defect, the only cure being to renew the valve, and if worn badly in corresponding areas, the carburettor body.
13 Examine the jet needle for scratches or wear along its length. If either the valve or needle is seen to be defective, then detach the valve from the throttle cable by grasping the valve firmly in one hand whilst compressing the return spring against the carburettor top with the other. Disengage the throttle cable from its retaining slot in the valve and place the valve, together with its return spring, on a clean work surface.
14 The jet needle may be detached from the throttle valve after removal of its retaining clip. This clip should be eased from position with either the flat of a small screwdriver or a pair of long-nosed pliers. Note the fitted position of the clip attached to the jet needle and then carefully push it from its retaining groove. This clip should normally be fitted in the second groove from the top of the needle. Check that the needle is not bent by rolling it on a flat surface, such as a sheet of plate glass. If in doubt as to the condition of the needle, or the jet through which it passes, return it to an official Honda service agent who will be able to give further advice and, if necessary, provide a new component.
15 Inspect the valve return spring for signs of fatigue, failure or severe corrosion and renew it if found necessary. The sealing ring located within the carburettor top must be renewed if seen to be perished or in any way damaged. The procedure adopted for reassembly of the throttle valve component parts should be a direct reversal of that used for dismantling.
16 Prior to reassembly of the carburettor, check that all the component parts, both new and old, are clean and laid out on a piece of clean rag or paper in a logical order. On no account use excessive force when reassembling the carburettor because it is easy to shear any one of the screw threads. Furthermore, the carburettor is cast in a zinc based alloy which itself does not have a high tensile strength. If any of the castings are damaged during reassembly, they will almost certainly have to be renewed.
17 Reassembly is basically a reversal of the dismantling procedure, whilst noting the following points. If in doubt as to the correct fitted position of a component part, refer either to the figure accompanying this text or to the appropriate photograph.
18 When fitting the throttle stop and pilot air screws, ensure that each screw is first screwed fully in, until it seats lightly, and then set to its previously noted position. Alternatively, set the pilot air screw 1¾ turns out from fully in, the setting of the throttle stop screw will then have to be determined by following the adjustment procedure listed in Section 9 of this Chapter.

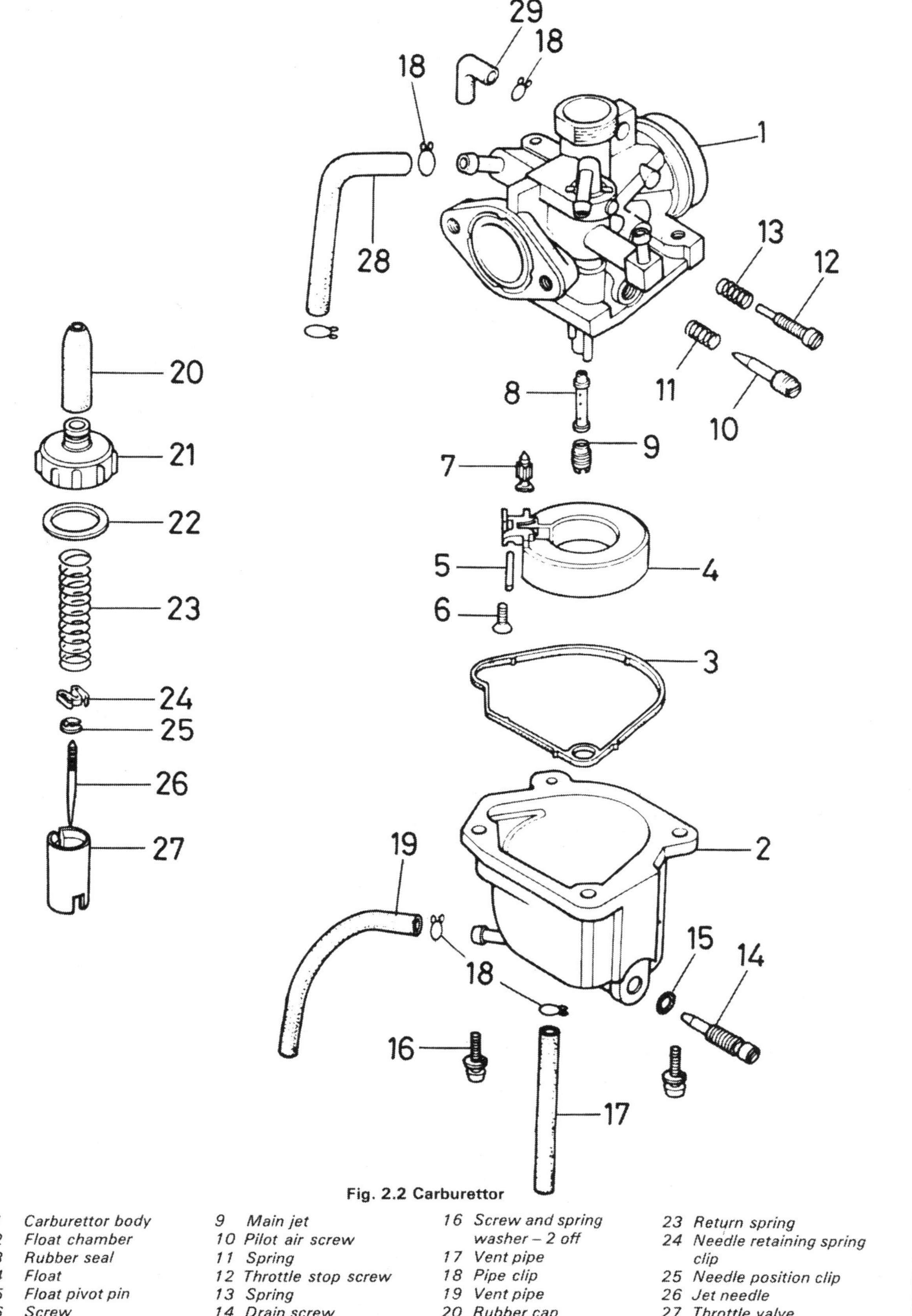

Fig. 2.2 Carburettor

1 Carburettor body
2 Float chamber
3 Rubber seal
4 Float
5 Float pivot pin
6 Screw
7 Float needle
8 Needle jet
9 Main jet
10 Pilot air screw
11 Spring
12 Throttle stop screw
13 Spring
14 Drain screw
15 O-ring
16 Screw and spring washer – 2 off
17 Vent pipe
18 Pipe clip
19 Vent pipe
20 Rubber cap
21 Carburettor top
22 Gasket
23 Return spring
24 Needle retaining spring clip
25 Needle position clip
26 Jet needle
27 Throttle valve
28 Fuel feed pipe
29 Breather pipe

8.6a Closely inspect the carburettor flange O-ring ...

8.6b ... and the float chamber seal

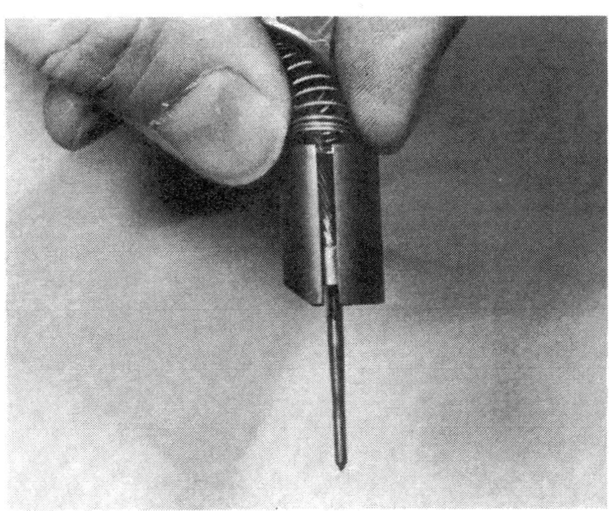
8.13 Detach the valve from the throttle cable ...

8.14a ... and remove the jet needle retaining clip

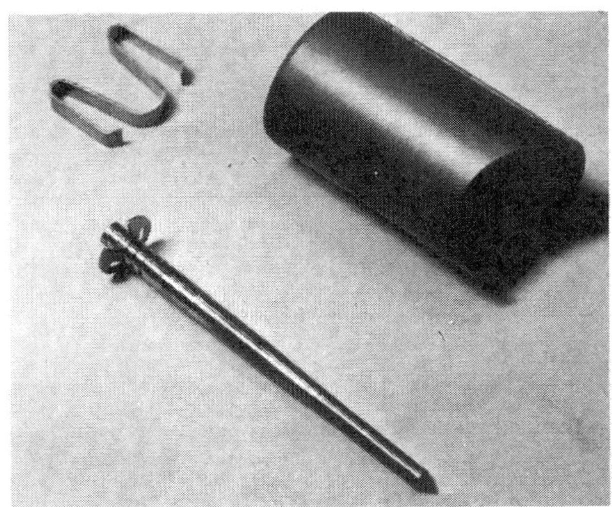
8.14b Inspect the throttle valve and jet needle

8.17a Relocate the float needle

8.17b Retain the float pivot pin in position with the cross head screw

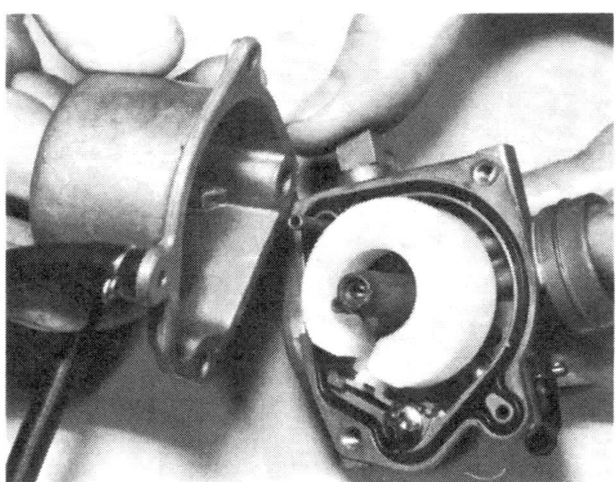
8.17c Refit the float chamber

8.17d Fit and tighten the drain screw

8.18a Fit and adjust the pilot air screw ...

8.18b ... followed by the throttle stop screw

9 Carburettor: adjustment

1 If flooding of the carburettor or excessive mixture weakness has been experienced, it is wise to start operations by checking the float level. This will involve detaching the carburettor, if not already removed, inverting it and removing the float chamber followed by the seal contained in the recess of the carburettor body mating surface.

2 To measure the float level, position the carburettor on a flat and level work surface so that its mouth is in contact with the surface; this will place the float in a vertical plane with its pivot point uppermost. The float level is correct if the distance between the mating surface of the carburettor body and the furthest point of the float is 12.2 ± 1.0 mm (0.48 ± 0.04 in) when the float valve is *just* closed. If the measurement taken shows the float level to be incorrect, then Honda recommend that the float assembly be renewed. Before going to the expense of buying a replacement item, try altering the float level by carefully bending the small tongue against which the float needle abuts.

3 The idle speed can be adjusted as follows. Run the engine until it reaches normal operating temperature, noting that this is

best done by riding the machine for a few miles. Identify the pilot air screw and the throttle stop screw. The heads of both of these screws may be seen through the forwardmost slot in the carburettor shield, the pilot air screw being the lower of the two.

4 Important note: before undertaking any adjustment work with the engine running, it is essential to remember that the Melody has an automatic transmission system. It will be appreciated that as the idle speed rises the centrifugal clutch will engage and the machine will attempt to move off. Be particularly wary of inadvertently 'blipping' the throttle. The machine should be placed squarely on its centre stand, checking that the rear wheel is raised clear of the ground, before work commences.

5 Start the engine and allow it to idle. Starting from the nominal setting of $1\frac{3}{4}$ turns out from fully in, turn the pilot air screw clockwise until the engine is heard to falter. Note the position of the screw at this point and then turn it anti-clockwise through the nominal setting position until the engine is once again heard to falter. Again note the position of the screw and then turn it clockwise to a point approximately mid-way between the two noted positions. By this method it is possible to establish the exact setting at which the engine idle speed is fastest and most regular; the setting of the pilot air screw at the point coinciding approximately with its recommended nominal setting. Once the pilot air screw is thus set, turn the throttle stop screw to give a reasonably slow but reliable idle speed.

6 Always guard against the possibility of incorrect carburettor adjustment which will result in a weak mixture. Two-stroke engines are very susceptible to this type of fault which will cause rapid overheating and often subsequent engine seizure. Changes in carburation leading to a weak mixture will occur if the air filter is removed or if the exhaust system is tampered with in any way. Above all, do not add oil to the fuel in the mistaken belief that it will aid lubrication. Adequate lubrication is provided by the throttle controlled oil pump.

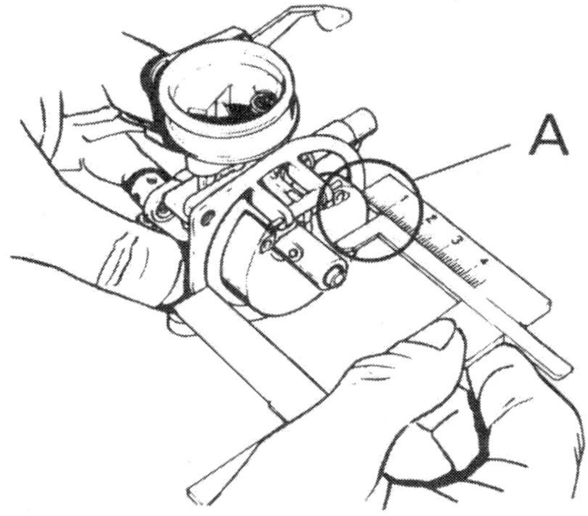

Fig. 2.3 Measuring the float level

A *Float level 12.2 ± 1.0 mm (0.48 ± 0.04 in)*

10 Carburettor: settings

1 The various jet sizes, the throttle valve cutaway and the jet needle position are all settings that are predetermined by the manufacturer. Under normal circumstances, it is unlikely that these settings will require modification, even though, in some cases, there is provision made. If a change appears necessary, it can often be attributed to a developing engine fault. The carburettor jets are pressed into position and should therefore be considered to form a permanent part of the carburettor body. No jet sizes are given; if a jet is suspected of being worn, the complete carburettor body should be returned to an official Honda service agent for further examination and, if necessary, renewal. Note that the recommended fitted position of the jet needle clip is in the second groove down from the top of the needle.

2 Before suspecting a carburettor fault, check that both the air filter assembly and the exhaust system are in good condition. Check also all other component parts in the fuel system, such as the vacuum-operated fuel tap and the fuel and vacuum pipes. With the float level set correctly, take the machine on to the road and ride it for a distance of approximately six miles. On completion of this ride, allow the engine to cool and then remove the spark plug. Inspect the plug insulator and electrodes for condition and colour in accordance with the instructions given in Chapter 3 of this Manual. Once a conclusion has been reached as to the fault, then take the steps necessary to obviate it whilst remembering that if the fault lies in the strength of the fuel/air mixture ratio, then it is necessary to err slightly on the side of richness. A weak mixture strength will cause the engine to overheat with potentially disastrous results.

3 Mixture strength may be altered by following the instructions given in the following table.

Component	Weaker mixture	Richer mixture
Pilot air screw	Turn screw anti-clockwise (out)	Turn screw clockwise (in)
Jet needle	Move clip towards top of needle to lower needle	Move clip towards tip of needle to raise needle

Note that a badly worn main jet will give an over rich mixture. The problem of jet renewal will have to be referred to a Honda service agent.

4 As an approximate guide, the pilot jet setting controls engine speed up to $\frac{1}{8}$ throttle. The throttle valve cutaway controls engine speed from $\frac{1}{8}$ to $\frac{3}{4}$ throttle. The size of the main jet is responsible for engine speed at the final $\frac{3}{4}$ to full throttle. It should be added however, that these are only guidelines. There is no clearly defined demarcation line due to a certain amount of overlap that occurs between the carburettor components involved.

11 Reed valve induction system: mode of operation

1 Of the various systems of controlling the induction cycle of a two-stroke engine, Honda has chosen to adopt the reed valve, a device which permits precise control of the incoming mixture, allowing more favourable port timing to give improved torque and power outputs. The reed valve assembly comprises a wedge-shaped die-cast aluminium alloy valve case mounted in the inlet tract. The valve case has rectangular ports which are closed off by flexible stainless steel reeds. The reeds seal against a heat and oil resistant synthetic rubber gasket which is bonded to the valve case. A specially shaped valve stopper, made from cold rolled stainless steel plate, controls the extent of movement of the valve reeds.

2 As the piston ascends in the cylinder, a partial vacuum is formed beneath the cylinder in the crankcase. This allows atmospheric pressure to force the valves open, and a charge of fuel/air mixture flows past the valve and into the crankcase. As the pressure differential becomes equalised, the valves close, and the incoming charge is then trapped. The charge of mixture in the cylinder is by this time fully compressed, and ignition takes place driving the piston downwards. The descending piston eventually uncovers the exhaust port, and the hot exhaust gases, still under a cetain amount of pressure, are

discharged into the exhaust system. At this stage, the charge of combustion mixture which has been compressed in the crankcase is released into the cylinder via the transfer ports, and the piston again ascends to close the various ports and begin compression. The reed valves open once more as another partial vacuum is created in the crankcase, and the cycle of induction thus repeats. It will be noted that no direct mechanical operation of the valve takes place, the pressure differential being the sole controlling factor.

12 Reed valve: removal, examination, renovation and refitting

1 The reed valve assembly is sandwiched between the base of the intake adaptor and the crankcase. In effect the reed valve is a flap valve which relies on atmospheric pressure to open it during the induction stroke, allowing the incoming mixture to flow into the crankcase, and close it when the crankcase pressure increases, to help prevent mixture being blown back through the carburettor.

2 Failure of the reed valve in service is not usual, although after a considerable amount of mileage has been covered the valve reeds may lose some springiness and their performance will suffer accordingly. Access to the reed valve assembly may be gained after removal of the following items:

The carburettor shield and centre body panel
The air filter housing
The carburettor
The exhaust pipe and silencer heat guards
The exhaust system
The air trunking and cylinder head cowling
The choke control valve
The intake adaptor

The carburettor shield is secured in position by a single retaining screw which passes through its forward edge. The air filter housing may be removed by unscrewing its single retaining bolt and pulling the housing rearwards to clear the carburettor mouth. The carburettor should be detached from the intake adaptor by removing its two retaining bolts. Disconnect the pipe running from the choke control valve to the intake adaptor and detach the choke control valve from the cylinder head by removing its two retaining screws. Manoeuvre the carburettor and the choke control valve rearwards away from the cylinder head.

3 Full information on removal and fitting of the exhaust system components is contained in Section 16 of this Chapter. With the exhaust system removed, detach the HT lead suppressor cap from the spark plug and remove the spark plug.

4 Removal of the air trunking and cylinder head cowling assembly is simply a matter of working in a logical sequence, commencing with the two cross head screws which retain the plastic grid in position and then detaching the trunking which should be followed in turn by each of the two cowling halves.

5 With all four of its retaining bolts thus exposed, the intake adaptor may be freed from the crankcase to expose the reed valve assembly located beneath it. If the valve frame is found to be stuck to the mating face of either the crankcase or intake adaptor, then use a thin blade to displace it.

6 Check the condition of the valve reeds and the curved valve stopper plate. Although the reeds and plate can be detached from the frame, to do so is pointless because the parts are not available as separate items. If damage is evident, in the form of cracking or distortion of the assembly, the whole unit must be renewed.

7 The sealing gasket fitted to each side of the valve frame must be in excellent condition. It is recommended that both of these gaskets be renewed every time the reed valve is disturbed.

8 The reed valve and its associated components should be refitted by reversing the dismantling procedure, whilst noting the following points. The four bolts which retain the intake adaptor in position on the crankcase should be tightened in even increments and in a diagonal sequence to a torque loading of 6.0 – 9.0 lbf ft (0.8 – 1.2 kgf m). Do not omit to refit the carburettor cover retaining bracket beneath the head of the forward outer bolt.

9 With the left-hand section of cowling located between the intake adaptor and the cylinder barrel, fit the right-hand section of cowling and secure both cowling halves in position by fitting and tightening the three retaining bolts. Having refitted the air trunking, spark plug and exhaust system, check that the O-ring attached to the mating flange of the carburettor is both serviceable and correctly located in its retaining groove before placing the carburettor against the intake adaptor. Fit and tighten the two carburettor retaining nuts, with bolts, to a torque loading of 7.0 – 9.0 lbf ft (0.9 – 1.2 kgf m).

10 With the choke control valve fitted to the cylinder head, reconnect the disturbed pipe to the intake adaptor and secure it in position with its spring clip. Refit the air filter housing and start the machine. Carry out a check for leaks at all of the fuel and oil pipe connections; although none of these connections should have been disturbed, it is possible that any one of them may have been placed under sone form of strain whilst the carburettor and intake adaptor were removed. Finally, refit the carburettor shield.

12.5 Displace the reed valve from the crankcase ...

12.6 ... and inspect the valve reeds and stopper plate

13 Air filter: removal, cleaning and refitting of the element

1 The air filter consists of a moulded plastic housing which contains an oil-impregnated foam element. Access to this element may be gained after removal of the carburettor shield and the housing cover. The carburettor shield is retained in position by a single cross head screw which passes through its forward edge. The housing cover is retained in position by two cross head screws whose heads are obscured by the outer face of the cover.

2 Once exposed, the filter element should be carefully eased out of its location in the filter housing. Closely examine the foam of the element for signs of damage, hardening or perishing caused by age. The element must be renewed if any one of these faults is apparent.

3 If the element is considered to be serviceable, detach the foam from its metal holder and clean it thoroughly in a non-flammable solvent, such as white spirit, whilst gently squeezing it to remove all traces of oil and accumulated dust. After cleaning, squeeze out the foam by pressing it between the palms of both hands and then allow a short time for any solvent remaining in the foam to evaporate. Do not wring out the foam as this will cause damage and thus lead to the need for early renewal.

4 Reimpregnate the foam with clean SAE 10W/30 oil and gently squeeze out any excess oil before refitting the foam in its holder. Refit the filter element into the filter housing and refit the housing cover. Note that great care should be taken when positioning both the element and cover to ensure that no incoming air is allowed to bypass the element or leak in past the cover joint. If this is allowed to happen, it will allow any dirt or dust that is normally retained by the element to find its way into the carburettor and crankcase assemblies; it will also effectively weaken the fuel/air mixture. Refit the carburettor shield.

5 Note that if the machine is being run in a particularly dusty or moist atmosphere, then it is advisable to increase the frequency of cleaning and reimpregnating the element. Never run the engine without the element fitted. This is because the carburettor is specially jetted to compensate for the addition of this component and the resulting weak mixture will cause overheating of the engine with the probable risk of severe engine damage.

13.1 Remove the filter cover retaining screws ...

13.2 ... and detach the cover to expose the air filter element

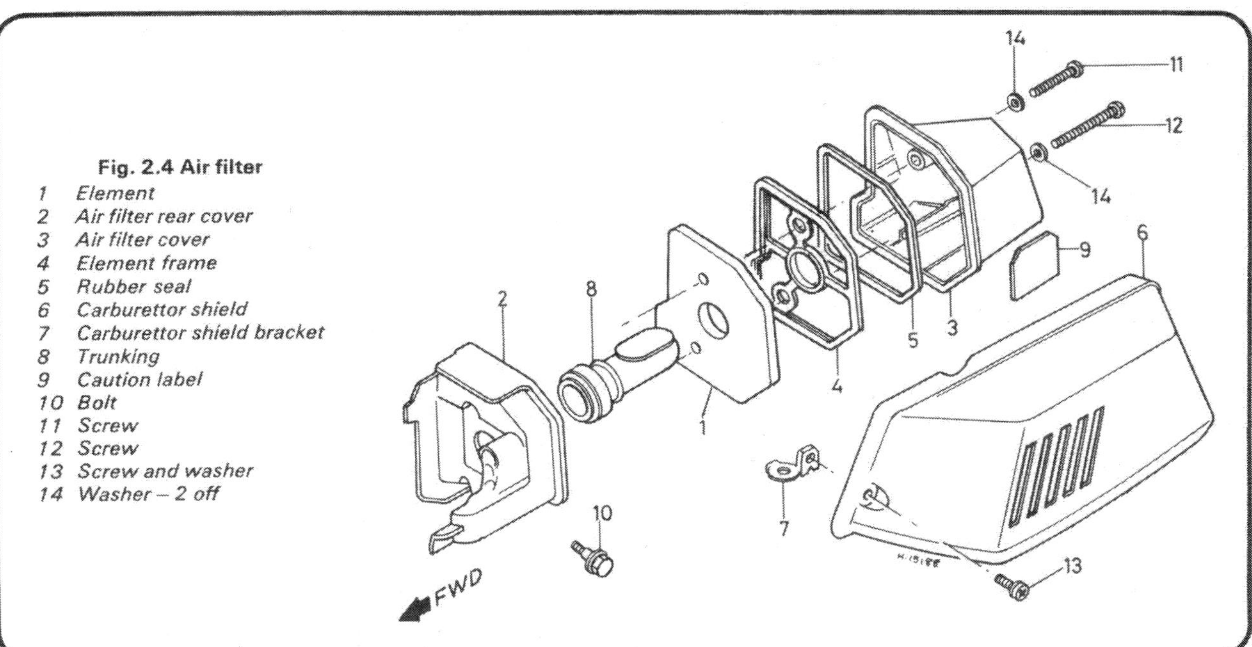

Fig. 2.4 Air filter

1 *Element*
2 *Air filter rear cover*
3 *Air filter cover*
4 *Element frame*
5 *Rubber seal*
6 *Carburettor shield*
7 *Carburettor shield bracket*
8 *Trunking*
9 *Caution label*
10 *Bolt*
11 *Screw*
12 *Screw*
13 *Screw and washer*
14 *Washer – 2 off*

14 Choke control filter: removal, cleaning and refitting of the element

1 The choke control filter consists of a moulded plastic base which forms a union for one of the pipes leading from the choke control valve to the carburettor. A rubber cap is fitted over this case and serves both as an air inlet and as a means of housing a small oil-impregnated foam element. Access to this element can be gained by removing the single screw which serves to retain the carburettor shield in position, detaching the shield and then carefully easing the rubber cap free of its plastic base.
2 Once removed, the filter element should be closely examined for signs of damage, hardening or perishing caused by age. The element must be renewed if any one of these faults is apparent. Check also that neither the filter cap not base are split or in any way damaged.
3 If the element is considered to be serviceable, then it should be cleaned thoroughly in a non-flammable solvent, such as white spirit, whilst gently squeezing it to remove all traces of oil and accumulated dust. After cleaning, squeeze out the foam by pressing it between two fingers and then allow a short time for any solvent remaining in the foam to evaporate. Do not wring out the foam as this will cause damage and thus lead to the need for early renewal.
4 Reimpregnate the foam with clean SAE 10W/30 oil and gently squeeze out any excess oil before placing it on a clean piece of paper or rag ready for refitting. Wipe clean the filter base and clean the air inlet passages within the filter cap by blowing them clear with a jet of compressed air. Wipe clean the inside of the filter cap and fit both the cap and element to the filter base. Finally, locate the carburettor shield in position and secure it by fitting and tightening its single retaining screw.

14.1 The choke control filter assembly

15 Exhaust system: decarbonisation and refinishing

1 The exhaust system is a one-piece welded unit comprising an exhaust pipe and silencer unit. It should be detached for cleaning if it is evident that restriction is causing a drop in engine performance. This condition is often displayed as an inability of the engine to reach maximum revolutions.

2 Due to the design of the exhaust system and in particular the complex internal substrate of the silencer box, cleaning by conventional means (dissolving deposits with chemicals) is not possible. It is recommended that all carbon deposits be removed from the front end of the exhaust pipe using a suitable scraper and if this fails to improve a blocked system, then the system should be renewed.
3 In skilled hands, it is possible to split open the exhaust system and burn off any deposits with a welding torch, and then weld up the system. Note, however, that this will have the disadvantage of ruining the paint finish and must only be carried out by someone experienced in this type of work. If the system is severely blocked, it advised that the advice of a Honda dealer is sought.
4 When new, the system is protected by a high temperature paint finish which prevents rusting, but in time this will begin to flake off leaving the silencer unprotected. If ignored, the thin sheet steel silencer box will quickly rust through, necessitating a new replacement item. This can be avoided by removing the assembly every six months or so. The silencer and pipe can then be cleaned of rust by using a wire brush and then degreased before having a fresh coat of heat resistant paint applied to their surfaces. One of the very high temperature (VHT) coatings such as Sperex is recommended.

16 Exhaust system: removal and refitting

1 Gain access to the forward section of exhaust pipe by removing the one dome head nut, with washer, and the single screw from the front of the centre body panel and then easing the panel clear of the machine. Remove the heat guard from the exhaust pipe by unscrewing its retaining bolt and retaining nut. Unscrew each of the two flange nuts which retain the exhaust pipe to the cylinder barrel. These nuts need only be unscrewed as far as the end of their studs. Remove the heat guard from the silencer by removing the single cross head screw which serves to retain it in position and then lifting the guard up off its locating clips.
2 Detach the exhaust system from the machine by removing the single bolt, with its plain washer, which passes through the mounting lug on the forward edge of the silencer. Grasp the system at either end and lift it clear of the cylinder barrel studs. Remove and discard the sealing ring which will be left imbedded in the recess of the exhaust port. This ring should be replaced with a new item before refitting the exhaust system.
3 Fitting of the exhaust system is a direct reversal of the removal sequence, whilst noting the following points. Make sure that the new sealing ring is located properly in the recess of the exhaust port before placing the flange of the exhaust pipe over it. Fit the silencer to crankcase mounting bolt, with washer, and tighten it, finger-tight. Tighten the two flange nuts which retain the exhaust pipe to the cylinder barrel and then fully tighten the silencer retaining bolt. Refit the heat guard to the silencer, the heat guard to the exhaust pipe and finally, the centre body panel.

17 Oil tank: removal and refitting

1 Commence removal of the oil tank by unscrewing the single flange nut which retains the rear mudguard to the underside of the rear body panel. This nut is located in the centre of the rear edge of the mudguard. With the nut removed, the mudguard can be eased down and rearwards to free it from its forward locating tabs.
2 Raise the seat of the machine and remove the battery from its housing by unscrewing its clamp retaining screw, lifting the clamp, removing the battery cover, disconnecting the battery leads and then carefully lifting the battery from position whilst

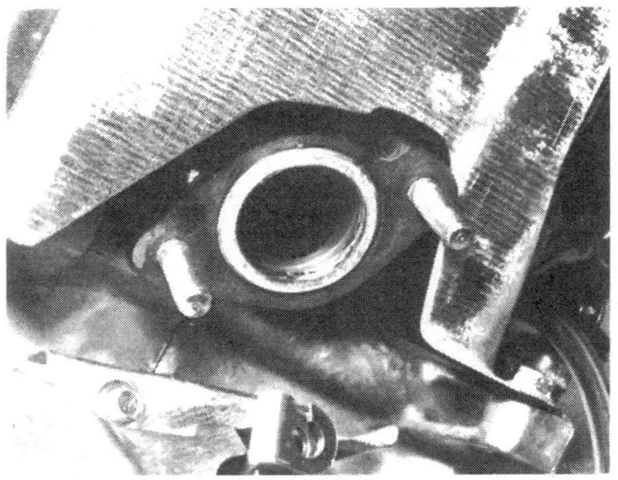
16.3a Locate the exhaust pipe sealing ring ...

16.3b ... before securing the exhaust pipe in position

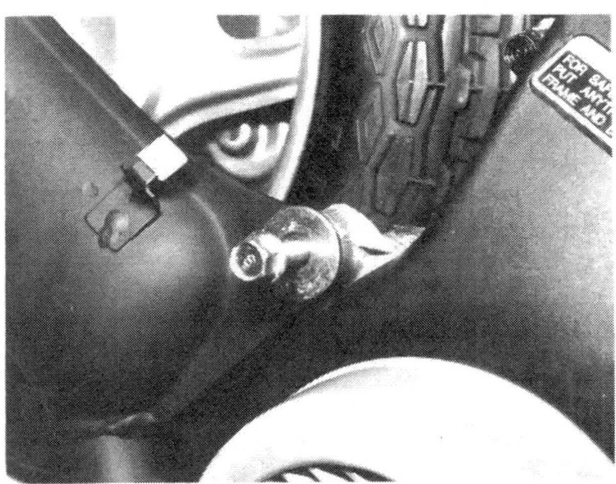
16.3c Fit the silencer to crankcase mounting bolt

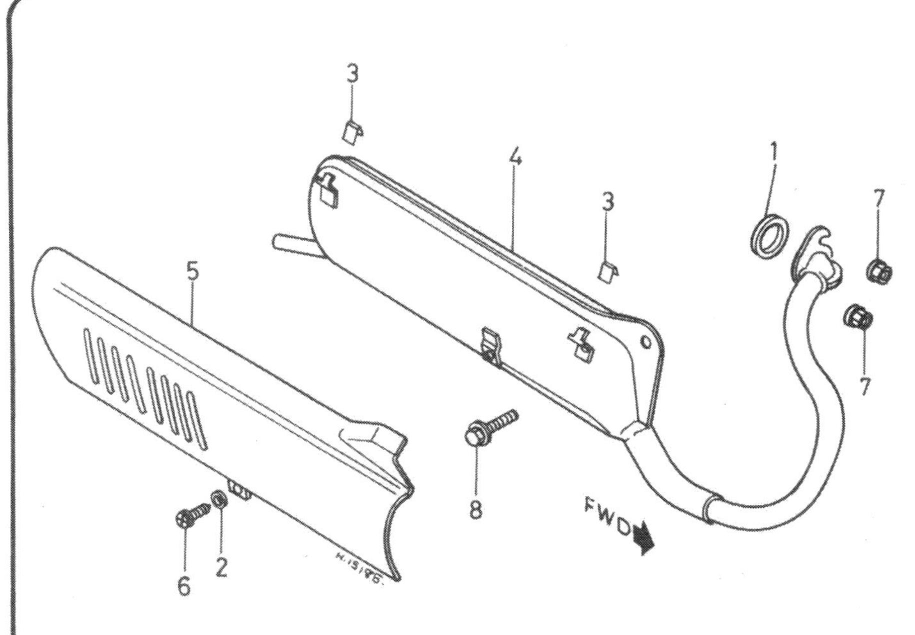

Fig. 2.5 Exhaust system

1 *Gasket*
2 *Damping rubber*
3 *Damping rubber – 2 off*
4 *Exhaust pipe*
5 *Heat guard*
6 *Screw and washer*
7 *Nut – 2 off*
8 *Bolt*

noting the location of its vent pipe in the housing. Remove the rubber pad from the base of the housing. Unscrew the battery holder retaining bolt and lift the holder from position.

3 Remove the luggage rack from its mounting points by unscrewing its two retaining bolts and two dome nuts with washers. Remove the two bolts which are located directly beneath the tail/stop lamp lens. Ease the lamp unit, complete with number plate, downwards to release it from its upper retaining clip. Pull the lamp unit away from the machine whilst carefully pulling the electrical wires out of their location in the body panel. Once the push connectors for these wires are exposed, note the colour coding of each wire and then pull apart the connectors to allow the lamp unit to be lifted away from the machine and placed in safe storage ready for refitting.

4 Remove the mounting bracket for the tail/stop lamp unit by unscrewing its two retaining bolts. Remove the oil tank filler cap and the small drip tray located beneath it. Carefully ease the rear body panel upwards over the oil tank filler point and then detach the battery vent pipe from the retaining stub on the underside of the battery housing. The rear body panel can now be lifted from position and placed in safe storage together with the lamp unit.

5 With the oil tank thus exposed, refit its filler cap and then obtain a clean container of at least 1.2 litre (2.1 Imp pint) capacity. Drain any oil from the tank by disconnecting the oil feed pipe from the tank to the oil pump at its connection with the pump and then placing the end of the pipe in the container. Note that it will be necessary to detach the centre body panel to gain full access to the oil pump. This panel is retained in position by a single dome head nut with washer and a single screw. Once the tank is drained, move to the right-hand side of the machine and trace the electrical wires from the oil level switch to the block connector located adjacent to the HT coil. Pull apart the two halves of this connector.

6 The oil tank can now be removed from the machine by unscrewing its single retaining bolt, which is located at the rear of the tank, and then carefully lifting it upwards and to the rear of the machine whilst ensuring that the oil feed pipe does not become caught on any engine or frame components. With the tank removed, now is the ideal time for removing and cleaning the oil filter in accordance with the instructions given in Section 18 of this Chapter.

7 Fitting the oil tank is a direct reversal of the removal procedure, whilst noting the following points. Do not omit to prime the oil feed pipe before reconnecting it to the oil pump. Full details of doing this and of bleeding the oil pump of air are given in Section 20 of this Chapter. Do not refit the centre body panel until the machine has been run and a comprehensive check for oil leaks carried out. Ensure that the level of oil in the tank is maintained at a reasonably high level throughout the bleeding procedure and that only oil of the correct type is used, that is, Honda 2-stroke injector oil or equivalent. With all electrical connections remade, check that the oil level switch is functioning correctly by unplugging it from the tank and moving the switch plunger to its 'On' and 'Off' positions whilst observing the warning light on the speedometer panel. Note that the battery should be temporarily reconnected and the ignition switch turned to the 'On' position before checking the switch. Check also that the tail and stop lamps are both functioning correctly before taking the machine onto the road.

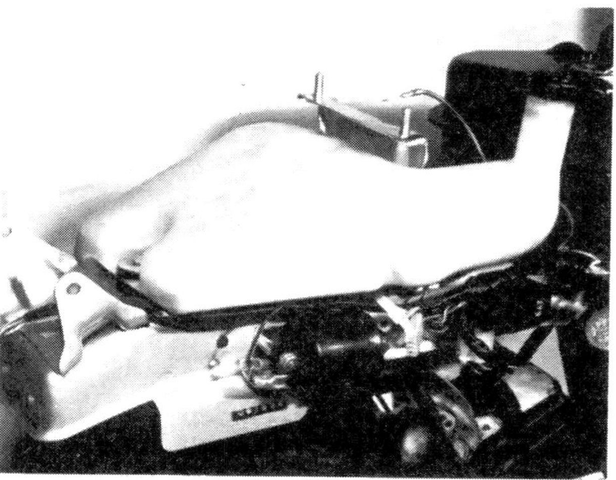

17.5 The exposed oil tank

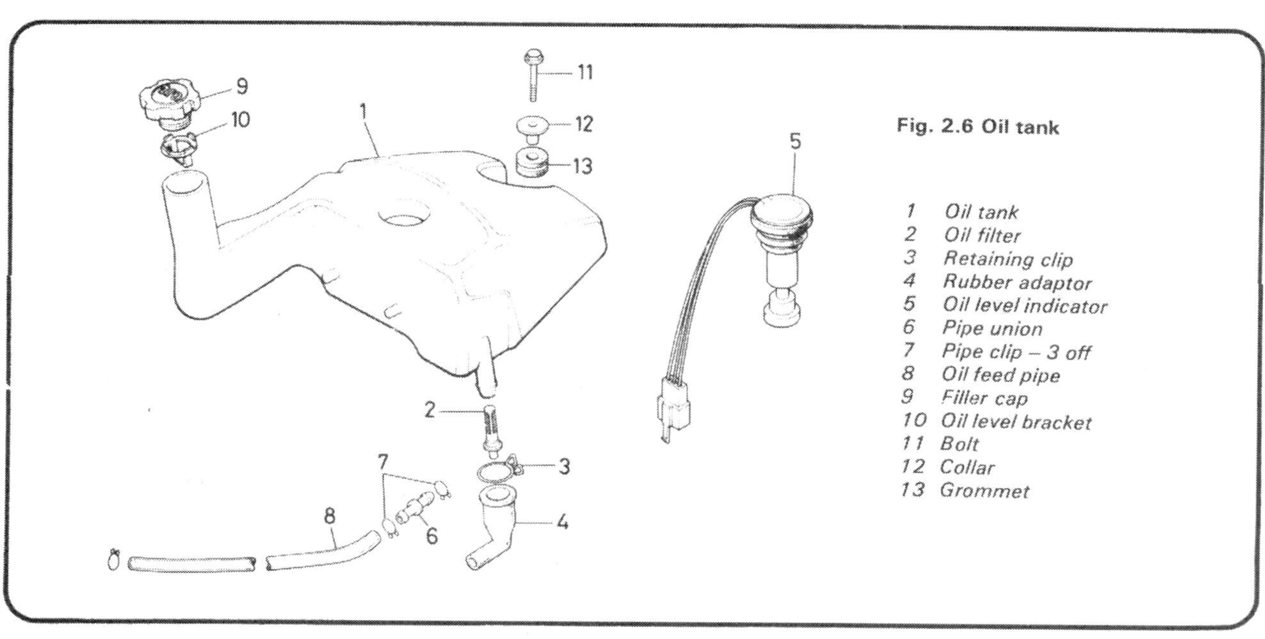

Fig. 2.6 Oil tank

1 *Oil tank*
2 *Oil filter*
3 *Retaining clip*
4 *Rubber adaptor*
5 *Oil level indicator*
6 *Pipe union*
7 *Pipe clip – 3 off*
8 *Oil feed pipe*
9 *Filler cap*
10 *Oil level bracket*
11 *Bolt*
12 *Collar*
13 *Grommet*

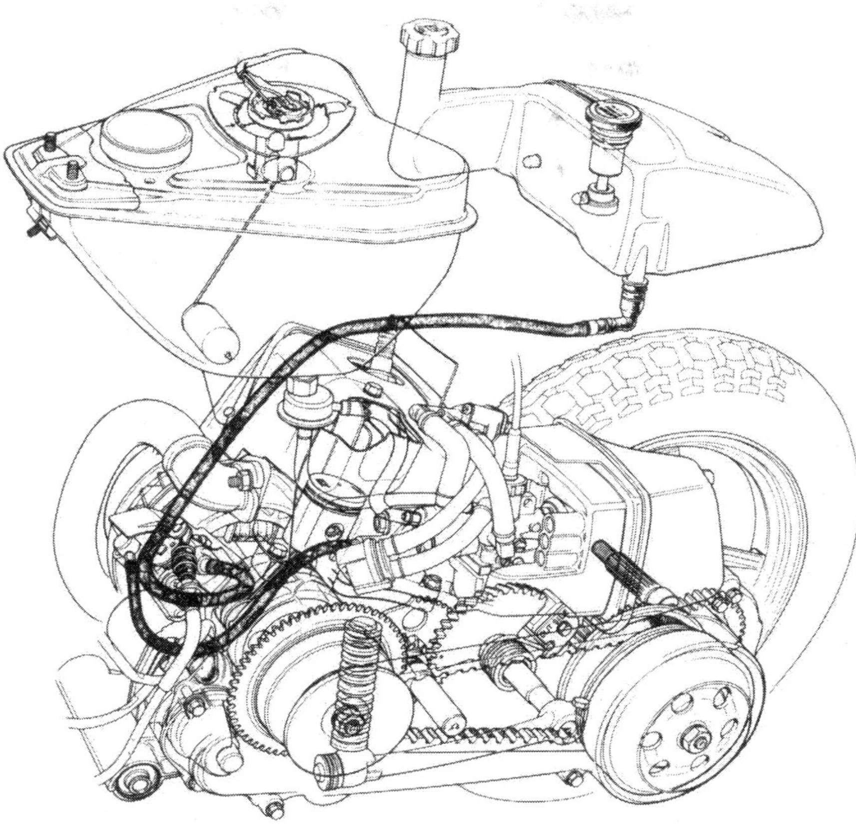

Fig. 2.7 Engine lubrication system

18 Oil filter: removal, cleaning and refitting

1 The oil filter comprises a small cylindrical filter element which is contained in a rubber adaptor which itself forms a union between the outlet point of the oil tank and the oil feed pipe to the oil pump. In order to gain access to this element it is first necessary to drain the tank of oil. To do this, obtain a clean container of at least 1.2 litres (2.1 Imp pints) capacity and place it beneath the machine. Remove the centre body panel from the machine by unscrewing its single dome head nut with washer and its single screw. This will give enough access to the oil pump to enable the oil feed pipe to be detached from the pump and its end placed in the container. On completion of draining the oil, unclip the rubber adaptor from the tank outlet and carefully lower the filter element clear of the tank. Note the routing of the pipe and then draw it clear of the machine.
2 If oil starvation problems have been experienced, yet the filter element is seen to be clean then the oil feed pipe must be checked for blockage or signs of damage leading to leakage of oil. Renew the pipe, if necessary, and proceed to clean the filter element as follows.
3 The element should be cleaned by directing a jet of compressed air through its open end as shown in the figure in the Routine maintenance section. Do not attempt to direct air through the sides of the element as this will only serve to collapse the fine filter mesh, making renewal of the element necessary. Take care to wear adequate eye protection against any spray back of oil from the element and remember to observe the necessary fire precautions. If necessary, any stubborn traces of contamination can be removed by gentle rubbing of the element with a soft-bristled brush; a used toothbrush is ideal. On completion of cleaning, closely inspect the element for any splits or holes that will allow the passage of sediment through it and into the pump. Renew the element if it is in any way defective.
4 Fitting of the filter components and oil feed pipe is a direct reversal of the removal procedure, whilst noting the following points. After having refilled the oil tank with Honda 2-stroke injector oil, or equivalent, prime the oil feed pipe and bleed the oil pump of air in accordance with the instructions given in Section 20 of this Chapter. Do not refit the centre body panel until the machine has been run and a comprehensive check for oil leaks carried out at all disturbed connections.

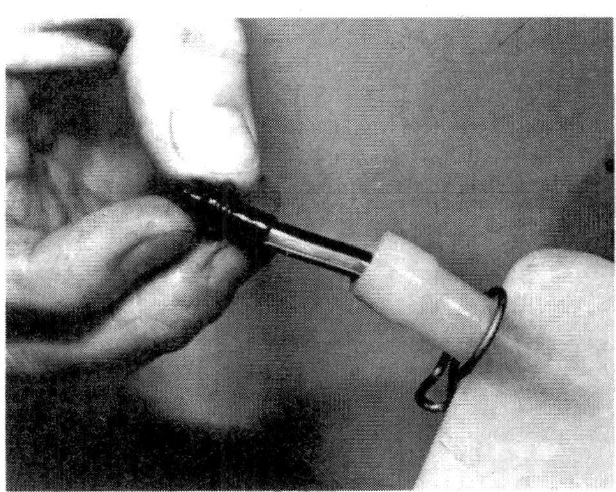

18.1 The oil filter element

19 Oil pump: removal, examination and refitting

1 Refer to Section 16 of this Chapter and remove the centre body panel together with the complete exhaust system from the machine.

2 Where the machine is equipped with a starter motor, move to the left-hand side of the machine and locate the electrical wires which are routed down the frame structure, just forward of the cylinder head. Pull apart the two halves of the only block connector to disconnect the starter motor from the main wiring harness. Remove the two bolts which retain the starter motor to the transmission casing.

3 It is now necessary to detach the engine from the frame and ease it far enough rearwards to allow access to be gained to the oil pump retaining bolt. Before doing this, obtain a block of wood, or similar, which will fit between the crankcase and floor, thus preventing the engine unit from dropping away from the frame directly it is detached.

4 Move to the rear wheel and unscrew the adjuster nut from the threaded end of the rear brake operating cable. Remove the cable anchor plate by unscrewing its single retaining bolt from the rear of the transmission casing. Check that the brake cable is free to slide through its guide at the front of the transmission casing. If the cable is clamped securely by the guide, then remove the guide retaining bolt.

5 Locate the single flange nut which retains the rear mudguard to the underside of the rear body panel. This nut is located in the centre of the rear edge of the mudguard. With the nut removed, the mudguard can be eased down and rearwards to free it from its forward locating tabs.

6 Unscrew the locknut from the end of the engine unit pivot bolt and remove it together with the plain washer. Using a soft-metal drift and a hammer, carefully drift the pivot bolt from position. With the bolt thus removed, carefully ease the engine unit far enough rearwards to allow full access to the oil pump retaining bolt. Rest the engine crankcase on the block positioned beneath it.

7 Where a starter motor is fitted, firmly grasp the body of the motor and pull it from position. It may be found that the motor has become quite firmly stuck in the transmission casing, in which case the flange of the motor body should be gently tapped with a soft-faced hammer to help free it. Do not attempt to lever the motor away from the casing as this will only cause damage to the mating surfaces.

8 Commence removal of the oil pump by tracing the oil feed pipe from the oil tank to the pump and releasing its retaining clip before pulling it clear of the pump union. Clip the end of this pipe to prevent any leakage of oil; a strong bulldog clip is ideal for this purpose. Alternatively, plug the end of the pipe with a bolt of the appropriate shank diameter.

9 Disconnect the oil outlet pipe from the pump by releasing its retaining clip before pulling it off its retaining stub. Allow any oil in the pipe to drain into a piece of absorbent rag or a small, clean container.

10 Disconnect the oil pump control cable from the pump lever by pushing the lever towards the cable and then releasing the end nipple from the lever. Release the cable adjuster from its retaining bracket by sliding the rubber gaiter off its threaded end and unscrewing the innermost of the two retaining nuts to allow the adjuster to be pulled outwards and clear of the bracket.

11 The oil pump is secured to the right-hand crankcase half by a single retaining bolt. Before removing this bolt and pulling the pump clear of the crankcase, clean the area of crankcase around the pump and the pump itself of any oil or road dirt. It is important to prevent contamination of this sort entering the crankcase through the pump location. Where a starter motor has been removed, take care to mask properly the motor location before cleaning commences.

12 The only parts of the pump assembly that should be removed are the pump mounting plate and the O-ring which forms a seal between the pump and crankcase. The pump itself is effectively a sealed unit as no replacement parts are listed by Honda. If the pump is defective, it should, therefore, be renewed as a complete unit.

13 Carry out a close inspection of the pump for signs of oil leakage and cracking of the pump casting. Check that the return spring of the pump lever is working efficiently and is not fatigued or broken. If the teeth of the pump drive shaft are seen to be broken or badly worn, then it must be assumed that the teeth of the crankshaft worm gear are in the same condition, as the two components will have worn in unison. It is possible to inspect the crankshaft worm gear through the pump location in the crankcase. It is necessary to make use of a mirror and a torch to obtain a clean view of the crankshaft. A defective worm gear will necessitate removal of the engine/transmission unit from the frame so that the crankcase halves may be separated to allow removal of the crankshaft assembly and renewal of the right-hand mainshaft.

14 Complete the inspection of the pump assembly by checking the condition of the O-ring fitted to the pump boss. This O-ring must be renewed if it is seen to be in any way damaged or has become flattened during use. Lightly lubricate both the O-ring and the teeth of the pump drive shaft with a lithium based grease before inserting the pump into the crankcase and carefully pushing it fully home. Fit and tighten the pump retaining bolt.

15 Where a starter motor is to be refitted, check that the O-ring fitted to the boss of the motor is both serviceable and correctly located in its groove before pushing the motor into its crankcase location. If the O-ring is dry, lubricate it with a little engine oil before fitting the motor. Align the motor with the transmission casing and fit and tighten its two securing bolts.

16 Reconnect the oil pump control cable to the pump lever and ease the engine unit forward to align with its frame mounting. Fit the pivot bolt and fit and tighten its locknut, together with its plain washer. Tighten this nut to a torque loading of 25 – 33 lbf ft (3.5 – 4.5 kgf m). Refit the rear mudguard and reroute the rear brake operating cable along the underside of the transmission casing. The cable must be located correctly in both its guide clamp and its anchor plate. Make sure that the clamp and plate retaining bolts are fully tightened. With the cable end passed through the trunnion of the brake cam operating lever, fit the adjuster nut to the cable end and tighten the nut until the amount of free play measured at the tip of the handlebar lever is 10 – 15 mm (0.4 – 0.6 in).

7 Reconnect the electrical leads of the starter motor to the main wiring harness. The oil pump control cable should now be checked for adjustment by opening the throttle as far as possible and then checking that the pointer on the pump lever is in exact alignment with the corresponding pointer cast in the pump body. Note that this check should only be carried out with the throttle cable correctly adjusted (see Section 7 of this Chapter). If cable adjustment is found to be incorrect, move the cable guide through its retaining bracket until both pointers come into alignment. Do this by rotating the guide retaining nuts in the required direction. With the pointers aligned, secure the guide in its retaining bracket by tightening the two nuts against each other.

18 Refer to Section 16 of this Chapter and refit the exhaust system. The oil pump should now be bled of all air in accordance with the instructions given in the following Section. On completion of this bleeding procedure, refit the centre body panel.

20 Oil pump: bleeding of air

1 If the oil pump has been removed, or the oil system drained or allowed to run dry, then it is most important to bleed any air from the system before the engine is started. Failure to do this will result in the engine seizing, with the resulting expense of a complete engine rebuild.

2 Before commencing with the pump bleeding procedure,

check that the pump control cable is in correct adjustment. Details of checking this adjustment are contained in the preceding Section of this Chapter. Check also that the oil feed pipes running to and from the pump are undamaged and show no signs of deterioration. Renew these pipes if necessary. Obtain a syringe or an oil container fitted with a long spout, either of which should be filled with clean engine oil. Check that the oil tank is full.

3 Commence the bleeding procedure by detaching the end of the oil feed pipe from its retaining stub on the oil pump and allowing oil to drain from the oil tank, through the pipe and into a piece of absorbent rag. Once the flow of oil from the pipe is seen to be free of air, pinch the end of the pipe together or plug it with a bolt of the appropriate shank diameter. Take the oil container and press its spout against the inlet stub of the oil pump. Pump as much oil into the pump as possible, unplug the end of the feed pipe and push it over the inlet stub. Retain the pipe in position with its spring clip.

4 With the oil pump and oil feed pipe thus primed with oil, it is now necessary to prime the oil outlet pipe which runs from the pump to the intake adaptor. Disconnect this pipe from the oil pump and raise its freed end above that which is connected to the intake adaptor. Using the oil container, fill the pipe with clean engine oil. Once the pipe is full, pinch its end together between two fingers and quickly reconnect it to the oil pump. Carefully use the protected jaws of a pair of long nose pliers to push the pipe fully onto the stub and then retain the pipe in position with its spring clip.

5 Start the engine and allow it to run at tick-over speed. Special attention should be paid to the presence of air in the outlet pipe from the oil pump directly the engine is started. Because of the method used to prime the outlet pipe, it is impossible to avoid a small amount of air getting into the pipe. In practice, this air was seen to take the form of a bubble which was approximately one tenth of an inch in length. This bubble gradually moved towards the intake adaptor during the pump bleeding procedure and was eventually expelled from the pipe to be followed by a constant flow of air-free oil from the pump. The presence of this small amount of air in the outlet pipe was considered to be acceptable.

6 Complete the oil pump bleeding procedure by holding the lever of the pump in its fully-open position with the tip of one finger whilst observing the oil outlet pipe at its union with the pump for any sign of air issueing from the pump. The presence of a constant flow of air from the pump will indicate the need for the pump bleeding procedure to be repeated. Allow the pump lever to return to its normal position directly all air has been expelled from the outlet pipe. Increasing the throttle opening for short intervals will help to move any air out of the outlet pipe but care must be taken not to race the engine unnecessarily. Carry out a final check for leaks at all disturbed connections before taking the machine onto the road.

19.11 Remove the oil pump retaining bolt

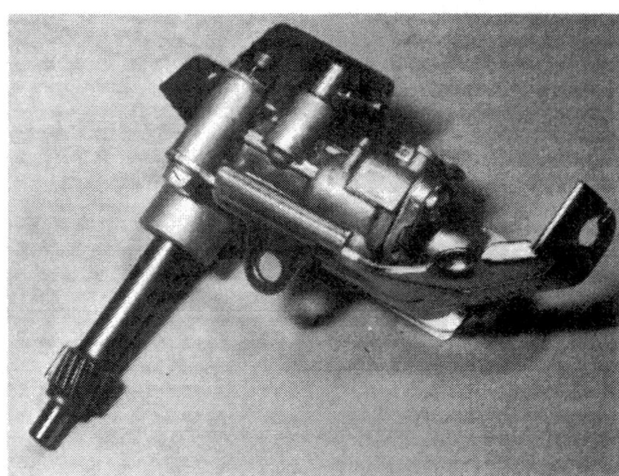

19.13 Inspect the teeth of the pump drive shaft ...

19.14 ... and check the condition of the O-ring fitted to the pump boss

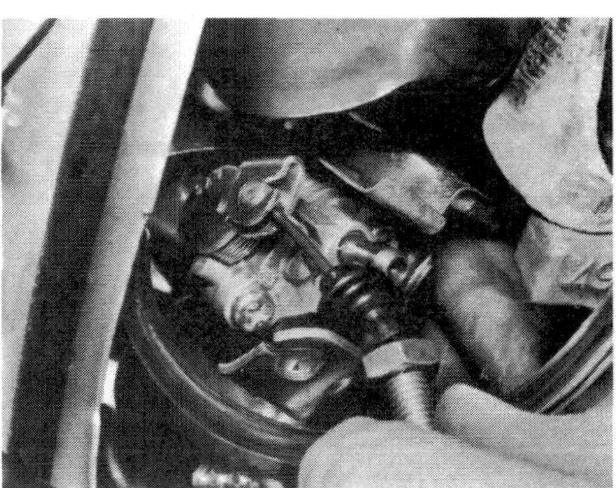

19.16 Connect the oil pump control cable to the pump lever

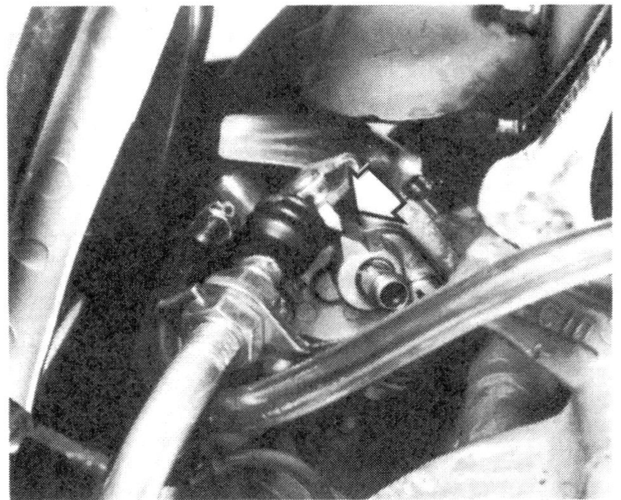

19.17a Align the two pointers as shown ...

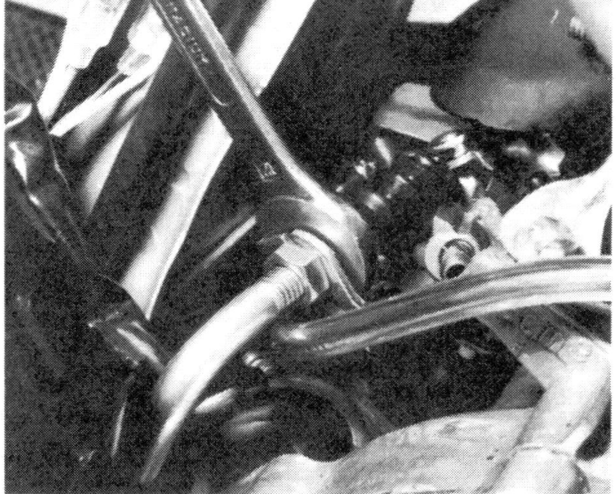

19.17b ... and secure the cable guide by tightening its securing nuts

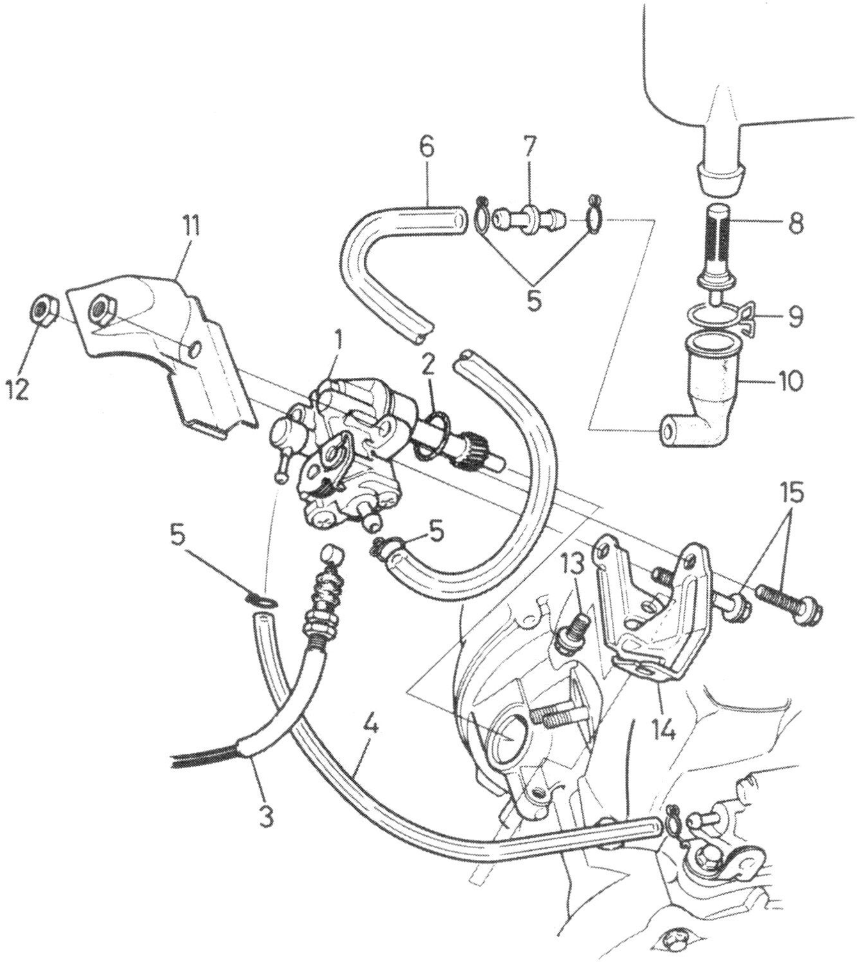

Fig. 2.8 Oil pump assembly

1 Oil pump	*6 Oil feed pipe*	*11 Oil pump cover*
2 O-ring	*7 Pipe union*	*12 Nut – 2 off*
3 Operating cable	*8 Oil filter*	*13 Bolt*
4 Oil delivery pipe	*9 Retaining clip*	*14 Mounting plate*
5 Pipe clip – 5 off	*10 Rubber adaptor*	*15 Bolt – 2 off*

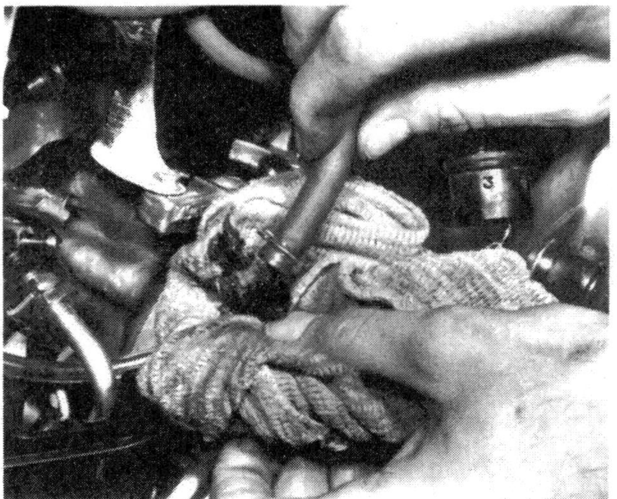

20.3a Allow oil to drain from the oil feed pipe

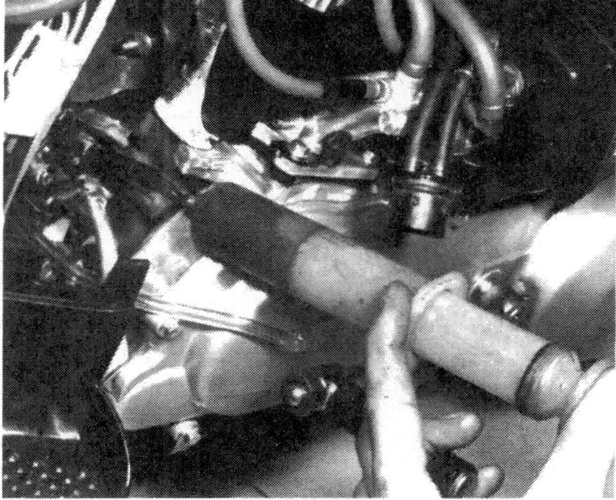

20.3b Prime the pump with oil and connect the feed pipe

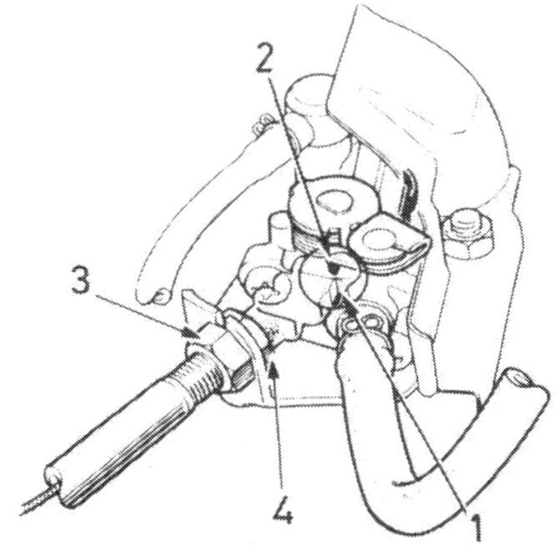

Fig. 2.9 Oil pump cable adjustment synchronisation marks

1 Oil pump alignment mark
2 Index mark
3 Adjusting nut
4 Locknut

20.4 Prime the outlet pipe before reconnecting it to the pump

Fault diagnosis: fuel system and lubrication, on following page

21 Fault diagnosis: fuel system

Symptom	Cause	Remedy
Excessive fuel consumption	Air cleaner choked or restricted	Clean or renew element
	Fuel leaking from carburettor	Check all unions and gaskets
	Badly worn or distorted carburettor	Renew
	Float sticking	Check and clean float needle seat and float pivot
	Jet needle setting too high	Adjust as described in text
	Main jet worn	Return carburettor to Honda service agent
	Carburettor flooding	Check float level Inspect float needle seat for wear
	Choke control valve defective	Check pipes leading to valve Check filter assembly Renew valve
Idling speed too high	Throttle stop screw in too far	Adjust screw
	Carburettor top loose	Tighten
	Throttle cable sticking or maladjusted	Disconnect and lubricate cable or renew. Adjust cable
	Choke control system defective	As above
Engine dies after running for a short while	Blocked air hole in fuel filler cap	Clean
	Dirt or water in carburettor	Remove carburettor and clean
	Blocked fuel tap filter	Remove tap and clean element
General lack of performance	Weak mixture: float needle sticking in seat	Remove float chamber and check needle seating
	Air leak at carburettor or leaking crankcase seals	Check for air leaks or worn seals
	Blocked air filter	Clean and relubricate element
	Choked exhaust system	Remove system and clean
	Choke control system defective	As above
Engine sluggish. Does not respond to throttle	Back pressure in exhaust system	Remove system and clean
	Air cleaner choked or restricted	Check and clean
	Choke control system defective	As above
	Fuel octane rating incorrect	Use correct grade (star rating) fuel

22 Fault diagnosis: lubrication system

Symptom	Cause	Remedy
White smoke from exhausts	Too much oil	Check oil pump setting and reduce if necessary
Engine runs hot and gets sluggish when warm	Too little oil	Check oil pump setting and increase if necessary Check oil supply to pump is not restricted by blocked filter or feed pipe Check oil outlet pipe for blockage Check for leakage of oil
Engine runs unevenly, not particularly responsive to throttle openings	Intermittent oil supply	Bleed oil pump to displace air in feed pipe Check for restriction in oil system Check for leakage of oil
Engine dries up and seizes	Complete lubrication failure	Check for blockage in oil system Check for failure of oil pump drive

Chapter 3 Ignition system

Refer to Chapter 7 for information on the NB, ND and NP50 models

Contents

Specifications

Ignition system

Type	CDI (capacitor discharge ignition)

Ignition coil

Primary winding resistance	0.2 – 0.3 ohm
Secondary winding resistance	3.4 – 4.2 K ohm

Ignition timing ... 15 ± 3° BTDC @ 2000 rpm

Spark plug

Make	NGK	Nippon Denso
Type	BPR4HS	W14FPR-L
Gap	0.6 – 0.7 mm (0.024 – 0.028 in)	

1 General description

The Honda Melody features what amounts to a maintenance-free electronic ignition system in which there are no moving parts and thus no mechanical wear. The CDI (capacitor discharge ignition) system comprises a solid state CDI unit and a magnetic pickup assembly. The latter replaces the traditional contact breaker assembly as the means of switching the ignition spark.

As the generator rotor rotates, current is generated in the ignition source coil which is transferred to and stored in the capacitor in the CDI unit. As the generator rotor moves further a trigger current is produced in the magnetic pick-up assembly which signals the stored power in the capacitor to discharge through the primary windings of the ignition coil. This surge of current induces a high voltage discharge from the ignition coil which is fed via the HT lead to produce the spark across the electrodes of the spark plug.

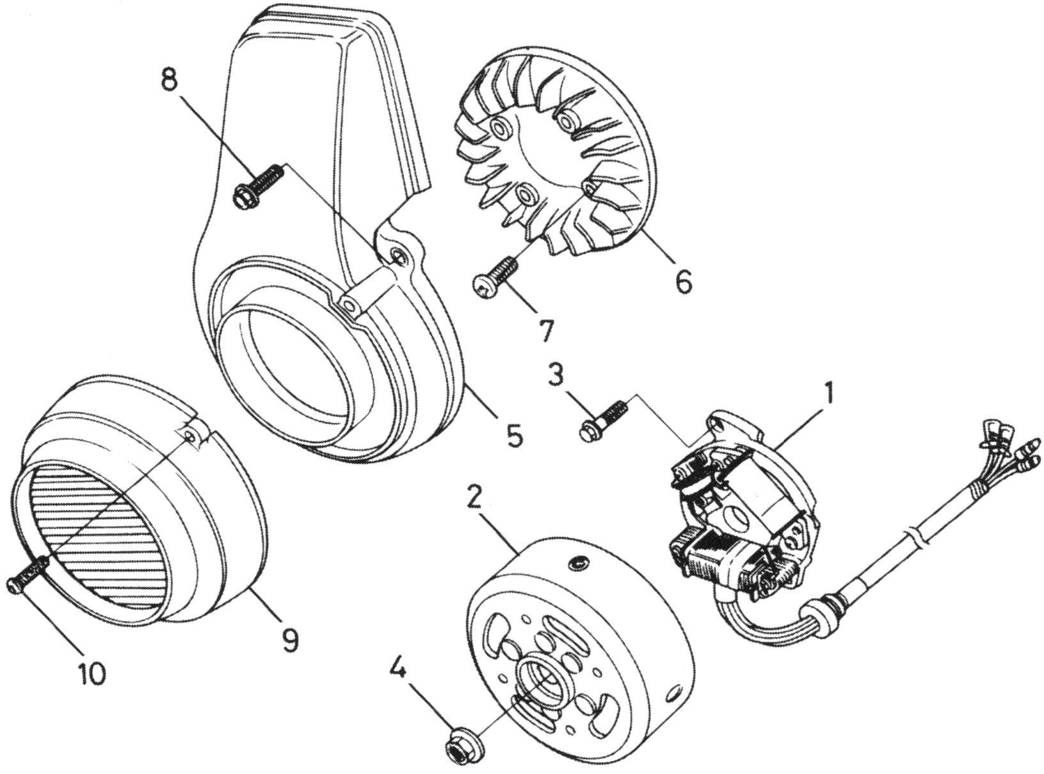

Fig. 3.1 Flywheel generator and cooling fan assembly

1 Stator
2 Flywheel
3 Bolt – 2 off
4 Nut
5 Fan cowling
6 Fan
7 Screw – 4 off
8 Bolt – 3 off
9 Plastic grid
10 Screw – 2 off

2 Checking the ignition system: general information

The CDI system is designed to require no regular maintenance, with the exception of the spark plug which should be cleaned and adjusted at the intervals specified in Routine maintenance. In the event that the system fails, it is likely that the fault lies in one of the associated wiring connections rather than the CDI unit or pickup assembly. Check the various terminals, connector blocks, wiring runs and the ignition switch before moving on to the major system components.

2 It should be noted that access to the pickup assembly and generator stator components will require the removal of the flywheel rotor. In order to achieve this, it was found necessary to obtain a good quality two-legged puller. It was also necessary to fabricate a simple tool with which to prevent the rotor from turning during tightening of its retaining nut. Full details of each tool are given in Chapter 1 of this Manual.

3 Great care must be taken when checking the ignition system to avoid electric shocks. The CDI unit produces high voltages which can prove to be very dangerous. It should be noted that the CDI unit, the pickup assembly and the ignition coil are all sealed units and in the event of a fault developing, repair is impracticable. Where a fault is suspected, it is recommended that the unit concerned is returned to a Honda service agent for final checking before a replacement item is purchased.

4 A limited range of tests is possible using a multimeter. These are fully described in the text of this Chapter for the benefit of owners who are suitably equipped and conversant with the use of this instrument. If in any doubt, then it is recommended that the assistance of a Honda service agent or a competent auto electrician is sought.

3 Checking the ignition timing

1 It cannot be overstressed that optimum performance of the engine depends on the accuracy with which the ignition timing is set. Even a small error can cause a marked reduction in performance and the possibility of engine damage as the result of overheating. To obviate the likelihood of the ignition timing moving from its correct setting, Honda have effectively fixed the position of the ignition pickup in relation to the flywheel rotor by designing the stator plate, upon which the pickup is mounted, so that it is bolted directly onto the engine crankcase without any provision for movement about its axis.

2 The ignition timing need only be checked if the flywheel generator assembly has been renewed or if it is suspected that the ignition system is functioning incorrectly. Expose the flywheel rotor by removing the two cross head screws which retain the gridded cover in position over the cooling fan and then detach the trunking from the cylinder head cowling. Locate the timing mark on the section of crankcase adjacent to the base of the cylinder barrel and the corresponding mark (F) on the wall of the rotor. Prepare each of these marks by degreasing them and then coating each one with a trace of white paint. This is not absolutely necessary, but will make the position of each mark far easier to observe if the light from the stroboscopic lamp to be used is weak or if the timing operation is to be carried out in bright conditions.

3 Full instructions on the use of a 'strobe' are given in the leaflet supplied by the manufacturer. The tool is used to observe the position of the timing marks whilst the engine is running. When the light from the lamp is aimed at the timing marks, it has the effect of 'freezing' the moving mark on the rotor in one position and thus the accuracy of the timing can be seen.

4 Two basic types of stroboscopic lamp are generally available, namely the neon and xenon tube types. Of the two, the neon type is much cheaper and will usually suffice if used in a shaded position, its light output being rather limited. The brighter but more expensive xenon types are preferable, if funds permit, because they produce a much clearer image.
5 Connect the 'strobe' to the HT lead, following the manufacturer's instructions. If an external 12 volt power source is required, a separate battery must be obtained, do not use the machine's battery or damage to the electrical system may result.
6 Start the engine and aim the 'strobe' at the timing marks. The ignition timing is correct when the mark on the crankcase is in alignment with the F mark on the rotor wall. This check should be carried out with the engine running at approximately 2000 rpm (just above idle speed) and a variance of 3 degrees either side of the specified setting may be allowed for.
7 If the timing marks are seen to be excessively out of alignment or if the mark on the rotor wall is seen to be 'wandering' either side of the crankcase mark, then it should be assumed that either the CDI unit or the flywheel generator assembly is defective and should be tested and, if necessary, renewed. Full instructions on testing each component part of the ignition system are contained in the following Sections of this Chapter. Note also the instructions given on tracing any wiring faults. One other point to remember is that in the unlikely event of the flywheel rotor being allowed to rotate about its crankshaft taper, due to the Woodruff key or its keyways having become excessively worn, the position of the timing marks will be seen to vary in relation to each other.

4 Checking the ignition system: tracing wiring faults

1 As mentioned, the single most likely cause of ignition failure is a damaged or broken wiring connection. It is assumed that the spark plug has been removed and checked by substitution to eliminate it as the source of the problem. Where possible, a multimeter set on the resistance position should be used to establish the likely location of the faulty connection. Failing this, connect up a dry battery and bulb as shown in the figure accompanying this text. The two probe leads can be fitted with small crocodile clips and can be connected to either end of a circuit to indicate continuity by lighting the bulb. This latter method will work just as well as a multimeter in this application, but whichever method is chosen, it is recommended that the machine's electrical system is isolated by disconnecting the battery.
2 The wiring should be checked in conjunction with the wiring diagram at the back of this Manual. Work in a logical sequence and include the ignition switch in the checks. Full details of checking this switch are included in Chapter 6 of this Manual.
3 Prior to testing each wiring run, visually examine all connections and terminals for signs of corrosion or arcing, cleaning or renewing each one as required. Check also that no wiring has been trapped between, or is fraying against, any moving frame or engine components. If no faults are discovered in the wiring runs and connections or in the ignition switch, then it can be assumed that the fault must lie in the CDI unit, the pickup assembly, the ignition coil or the HT lead and suppressor cap. Check also that power is being supplied to the ignition system by the ignition source coil.

5 Ignition source coil: testing and renewal

1 The ignition source coil is mounted on the stator plate of the flywheel generator assembly and is instrumental in creating the power in the ignition system. If it is suspected that failure of this coil has occurred, then it may be quickly and easily checked by carrying out the following instructions.
2 Remove the centre body panel from the machine to reveal the electrical wires which are routed down the left-hand side of the frame structure, just forward of the cylinder head. Unplug the push connector of the Black/red wire. Set a multimeter to its resistance function (x 1 ohm) and connect one of its probes to the Black/red wire and the other to a good earth point on the crankcase. If the reading shown by the multimeter is 50 – 300 ohm, then the coil is serviceable.
3 If the coil is found to be unserviceable, then the complete flywheel generator assembly will have to be renewed as Honda do not list its component parts as separate items. Always have a Honda service agent confirm that the coil is defective before going to the expense of purchasing a replacement item. Remember that various alternatives to purchasing a new item are available; these include obtaining advice from a competent auto-electrician as to whether the coil can be repaired and contacting the various 'bike breakers' who advertise in the weekly and monthly motorcycle journals. Removal and refitting of the flywheel generator assembly are dealt with in the relevant Sections of Chapter 1.

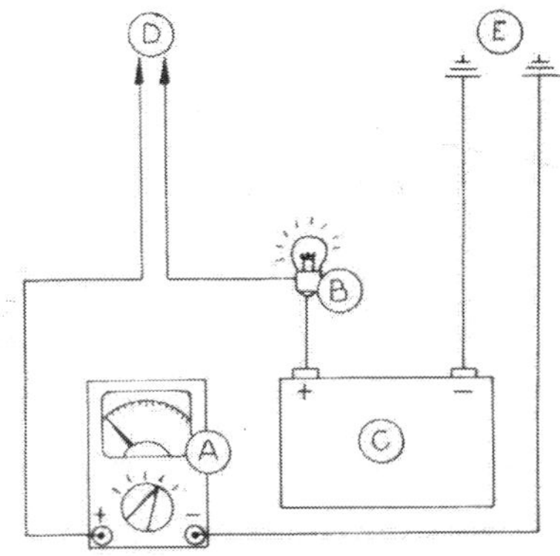

Fig. 3.2 Method of checking the wiring

A Multimeter
B Bulb
C Battery
D Positive probe
E Negative probe

5.1 Both the ignition source and the ignition pickup coils are mounted on the generator stator plate

6 Ignition pick-up coil: testing and renewal

1 The ignition pickup assembly is mounted on the stator plate of the flywheel generator assembly and serves as a means of switching the ignition spark. If it is suspected that failure of the pickup coil has occurred, then it may be quickly and easily checked by carrying out the following test procedure.
2 Remove the centre body panel from the machine to reveal the electrical wires which are routed down the left-hand side of the frame structure, just forward of the cylinder head. Unplug the push connector of the Blue/yellow wire. Set a multimeter to its resistance function (x 1 ohm) and connect one of its probes to the Blue/yellow wire and the other to a good earth point on the crankcase. If the reading shown on the meter scale is 10 – 100 ohm, then the coil is serviceable.
3 If the coil is found to be unserviceable, then the complete flywheel generator assembly will have to be renewed as Honda do not list its component parts as separate items. Always have a Honda service agent confirm that the coil is defective before going to the expense of purchasing a replacement item. Remember that various alternatives to purchasing a new item are available; these include obtaining advice from a competent auto-electrician as to whether the coil can be repaired and contacting the various 'bike breakers' who advertise in the weekly and monthly motorcycle journals. Removal and refitting of the flywheel generator assembly are dealt with in the relevant Sections of Chapter 1.

7 CDI unit: location and testing

1 The CDI unit is located on the right-hand side of the frame structure, immediately beneath the oil tank filler point. Removal of the centre body panel will give partial access to the unit; this panel is retained in position by one dome head nut with washer and a single screw. Detach the unit by unplugging its block connector and then easing the unit sideways out of its mounting sleeve.
2 The condition of the CDI unit can be determined by resistance testing in conjunction with the table of expected resistance readings for the different meter connections. It should be noted that the manufacturer recommends the use of specific meters for this test otherwise false readings may result. The meters are Sanwa type SP-10D (Part No 07308-0020000), or the Kowa meter TH-5H-2. If any other meter is used and the readings obtained conform with those given in the table it would be safe to assume that the CDI unit was functioning correctly. The individual terminals within the unit block connection are all clearly marked on the figure accompanying this text. The table clearly indicates which meter probe should be connected to which terminal and the limits between which the meter needle should fall if the unit is serviceable. Note that each measurement of resistance shown is quoted in kilo ohms.
3 If any one of the resistances measured in the test is outside the given limits, then the CDI unit must be replaced with a new item. If in doubt, ask a Honda service agent to verify the condition of the unit before going ahead with the purchase of a new item.

8 Ignition coil: location and testing

1 The ignition or high tension (HT) coil is located on the right-hand side of the frame structure immediately beneath the oil tank and is shielded by the rear body panel. It should not be necessary to remove the body panel in order either to disconnect the low tension (LT) and HT wires from the coil or to effect removal of the coil. It is far easier to cope with the awkwardness of working with the panel in place than to carry out the process of removing the panel.
2 The ignition coil serves to convert the LT pulse transmitted from the CDI unit into the HT pulse necessary to jump the gap between the spark plug electrodes and cause ignition of the fuel/air mixture being compressed in the combustion chamber. The pulse from the CDI unit creates a magnetic field of brief duration in the primary windings of the coil which in turn induces an HT pulse in the secondary windings which is then fed to the spark plug via the HT lead and suppressor cap.
3 To test the ignition coil, it is necessary to measure the resistance of both the primary and secondary windings. This can be done by using a multimeter that has both an ohms and kilo ohms scale. Test the primary winding by detaching the LT wire (Black/yellow) from the rear of the coil and then connecting one of the meter probes to the LT terminal on the coil whilst holding the other probe on a good earth point, preferably the laminated steel core of the coil. If the primary winding is serviceable, the meter reading will be 0.2 – 0.3 ohm. Any other reading will mean that the ignition coil is defective and in need of renewal.
4 Disconnect the HT lead from the coil by unscrewing its retaining cap. Alternatively, where the lead is permanently attached to the coil, remove the suppressor cap. Set the multimeter to its kilo ohms scale and place one of its probes on the lead terminal whilst holding the other probe on a good earth point. If the secondary winding is serviceable, the meter reading will be 3.4 – 4.2 K ohm. Any other reading will mean that the ignition coil is defective.
5 A defective ignition coil can be removed by detaching its LT wire and pulling the suppressor cap off the spark plug before unscrewing the two bolts that serve to retain the coil to the frame structure. Take care to retain the two spacer tubes through which the bolts pass and use a reversal of this removal procedure to fit the replacement coil. Note that if any doubt exists as to the condition of the coil, then it should be returned to a Honda service agent who will be able to confirm any suspected defect in the coil windings before supplying a new item. Remember that where the HT lead is permanently attached to the coil, any break in the core of the lead could well effect the meter reading whilst checking the secondary winding. If this proved to be the case, then the coil will have to be removed from the machine and taken to a Honda service agent or a competent auto-electrician to have the defective lead removed and a new one soldered to the coil terminal. This is a job that requires some degree of skill if the coil is not to be permanently damaged.

9 High tension lead: examination

1 Erratic running faults and problems with the engine suddenly cutting out in wet weather can often be attributed to leakage from the high tension lead and spark plug cap. If this fault is present, it will often be possible to see tiny sparks around the lead and cap at night. Once cause of this problem is the accumulation of mud and road grime around the lead, and the first thing to check is that the lead and cap are clean. It is often possible to cure the problem by cleaning the components and sealing them with an aerosol ignition sealer, which will leave an insulating coating on both components.
2 Water dispersant sprays are also highly recommended where the system has become swamped with water. Both these products are easily obtainable at most garages and accessory shops. Occasionally, the suppressor cap or the lead itself may break down internally. If this is suspected, the components should be renewed.
3 Where the HT lead is permanently attached to the ignition coil, it is recommended that the renewal of the HT lead is entrusted to an auto-electrican who will have the expertise to solder on a new lead without damaging the coil windings.

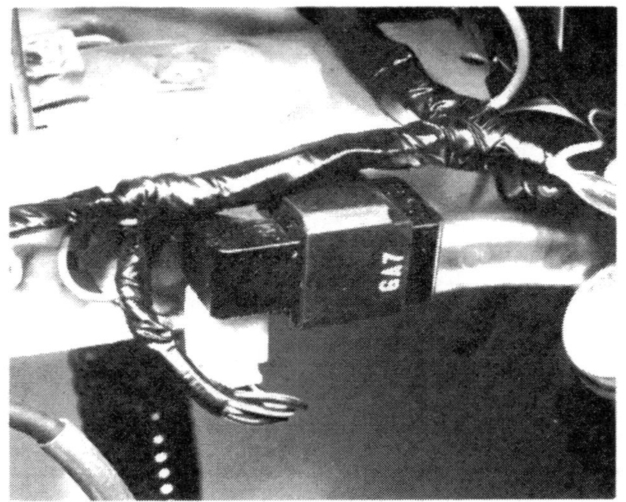
7.1 The CDI unit is located beneath the oil tank filler point

8.1 The ignition coil is located immediately beneath the oil tank

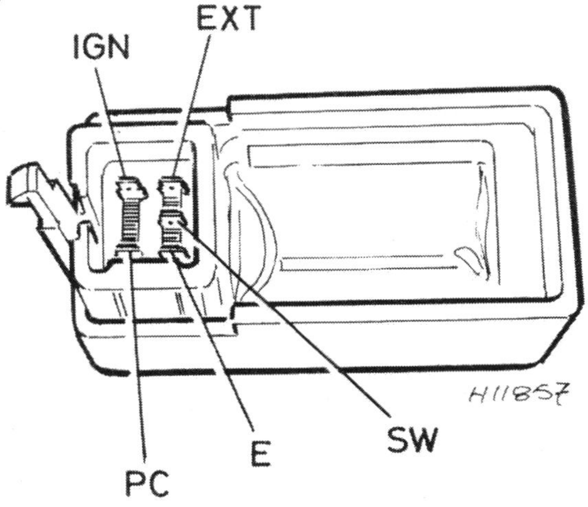

IGN
EXT
PC
E
SW
H11857

Meter range:
Sanwa – K ohm
Kowa – x 100 ohm

+ PROBE / – PROBE	SW	EXT	PC	E	IGN
SW		∞	∞	∞	∞
EXT	0.1– 10		∞	∞	"Needle swings then returns" or ∞
PC	0.5–200	0.5–50		1–50	∞
E	0.2– 30	0.1–10	∞		∞
IGN	∞	∞	∞	∞	

Fig. 3.3 CDI unit testing

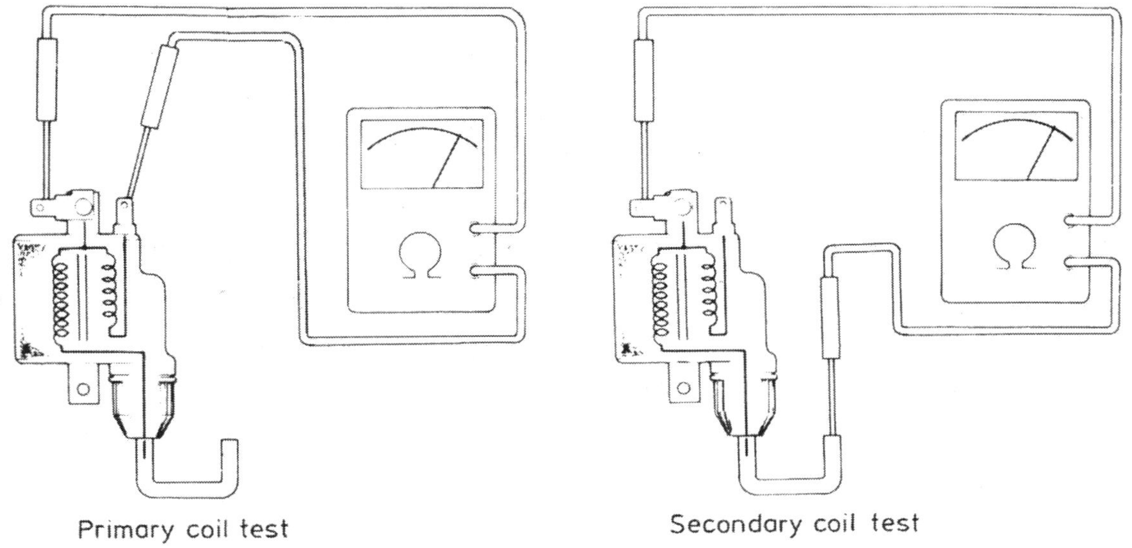

Primary coil test

Secondary coil test

Fig. 3.4 Ignition coil resistance tests

10 Spark plug: cleaning, checking and resetting the gap

1 Honda fit an NGK BPR4HS (or ND W14FPR-L) spark plug as standard equipment to the Melody. The recommended gap between the plug electrodes is 0.6 – 0.7 mm (0.024 – 0.028 in). The plug should be cleaned and the gap checked and reset at the service interval recommended in the Routine Maintenance Chapter at the beginning of this Manual. In addition, in the event of a roadside breakdown where the engine has mysteriously 'died' the spark plug should be the first item checked.

2 The plug should be cleaned thoroughly by using one of the following methods. The most efficient method of cleaning the electrodes is by using a grit blasting machine. It is quite possible that a local garage or motorcycle dealer has one of these machines installed on the premises and will be willing to clean any plugs for a nominal fee. Remember, before fitting a plug cleaned by this method, to ensure that there is none of the blasting medium left impacted between the porcelain insulator and the plug body. An alternative method of cleaning the plug electrodes is to use a small brass-wire brush. Most motorcycle dealers sell such brushes which are designed specifically for this purpose. Any stubborn deposits of hard carbon may be removed by judicious scraping with a pocket knife. Take great care not to chip the porcelain insulator round the centre electrode whilst doing this. Ensure that the electrode faces are clean by passing a small fine file between them; alternatively, use emery paper but make sure that all traces of the abrasive material are removed from the plug on completion of cleaning.

3 To reset the gap between the plug electrodes, bend the outer electrode away from or closer to the central electrode and check that a feeler gauge of the correct size can be inserted between the electrodes. The gauge should be a light sliding fit. Never bend the central electrode or the insulator will crack, causing engine damage if the particles fall in whilst the engine is running.

4 With some experience, the condition of the sparking plug electrodes and insulator can be used as a reliable guide to engine operating conditions. See the accompanying colour photographs.

5 Always carry a spare spark plug of the correct type. The plug in a two-stroke engine leads a particularly hard life and is liable to fail more readily than when fitted to a four-stroke.

6 Beware of overtightening the spark plug, otherwise there is risk of stripping the threads from the aluminium alloy cylinder head. The plug should be sufficiently tight to seat firmly on its sealing washer, and no more. Use a spanner which is a good fit to prevent the spanner from slipping and breaking the insulator.

7 If the threads in the cylinder head strip as a result of overtightening the spark plug, it is possible to reclaim the head by the use of a Helicoil thread insert. This is a cheap and convenient method of replacing the threads; most motorcycle dealers operate a service of this nature at an economic price.

8 Before fitting the spark plug in the cylinder head, coat its threads sparingly with a graphited grease. This will prevent the plug from becoming seized in the head and therefore aid future removal.

9 When reconnecting the suppressor cap to the plug, make sure that the cap is a good, firm fit and is in good condition; renew its rubber seals if they are in any way damaged or perished. The cap contains the suppressor that eliminates both radio and TV interference.

11 Fault diagnosis: ignition system

Symptom	Cause	Remedy
Engine will not start	Faulty ignition switch	Operate switch several times in case contacts are dirty. If lights and other electrics function, switch may need renewal
	Faulty CDI unit	Have unit tested. Replace if required
	Wiring fault	Check and repair wiring
	Faulty ignition source coil	Test and renew flywheel generator if necessary
	Faulty ignition pickup	As above
	Faulty HT coil	Test and replace if necessary
	Spark plug faulty or oiled up	Remove and check by substitution. Clean and regap plug
	Faulty HT lead or suppressor cap	Check by substitution
	Starter motor not working	Discharged battery. Remove battery from machine and recharge Faulty starter circuit. Check for continuity
	Short circuit in wiring	Check whether fuse is intact. Eliminate fault before switching on again
	Completely discharged battery	If lights do not work, remove battery and recharge
Engine misfires	Faulty ignition coil	Test and renew if required
	Wiring fault	Check and repair wiring
	Faulty HT lead	Renew lead
	Faulty spark plug	Renew plug or clean if fouled
	Faulty suppressor cap	Renew cap
	Faulty pickup coil	Test and renew if necessary
	Carburation fault	Refer to Chapter 2
Engine fades when under load	Pre-ignition	Check grade of plugs fitted; use recommended grades only

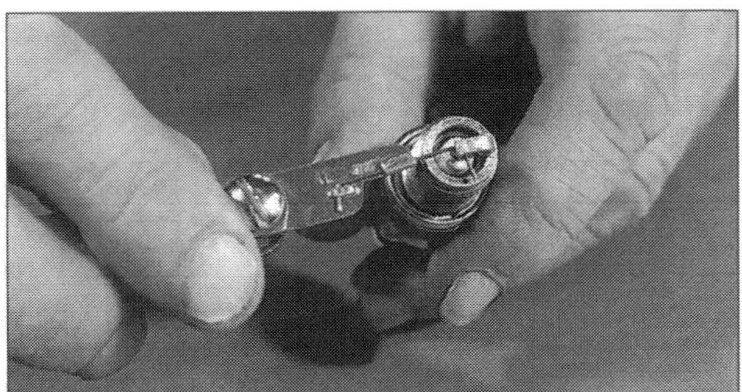

Electrode gap check – use a wire type gauge for best results.

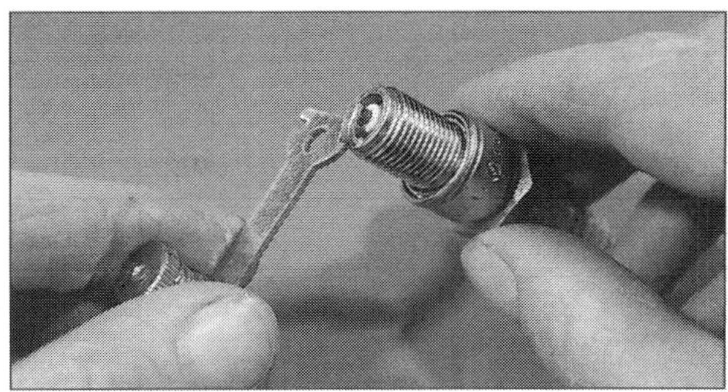

Electrode gap adjustment – bend the side electrode using the correct tool.

Normal condition – A brown, tan or grey firing end indicates that the engine is in good condition and that the plug type is correct.

Ash deposits – Light brown deposits encrusted on the electrodes and insulator, leading to misfire and hesitation. Caused by excessive amounts of oil in the combustion chamber or poor quality fuel/oil.

Carbon fouling – Dry, black sooty deposits leading to misfire and weak spark. Caused by an over-rich fuel/air mixture, faulty choke operation or blocked air filter.

Oil fouling – Wet oily deposits leading to misfire and weak spark. Caused by oil leakage past piston rings or valve guides (4-stroke engine), or excess lubricant (2-stroke engine).

Overheating – A blistered white insulator and glazed electrodes. Caused by ignition system fault, incorrect fuel, or cooling system fault.

Worn plug – Worn electrodes will cause poor starting in damp or cold conditions and will also waste fuel.

Chapter 4 Frame and forks

Refer to Chapter 7 for information on the NB, ND and NP50 models

Contents

Specifications

Frame

Type	Large diameter tubular spine welded to pressed steel rear section

Forks

Type	Leading link, undamped
Travel	55 mm (2.16 in)
Spring free length	117 mm (4.61 in)
Service limit	112 mm (4.41 in)
Shaft runout service limit	0.2 mm (0.008 in)

Rear suspension

Type	Pivoted engine/transmission unit supported by single suspension unit
Suspension unit	Coil spring and damper, non-adjustable
Spring free length	181.4 mm (7.14 in)
Service limit	175.5 mm (6.91 in)
Suspension travel	51 mm (2.01 in)

Torque wrench settings

	lbf ft	kgf m
Steering stem retaining nut	43 - 65	6.0 - 9.0
Front fork pivot arm nuts	14 - 22	2.0 - 3.0
Rear suspension unit:		
Lower bolt	18 - 25	2.5 - 3.5
Upper nut	22 - 29	3.0 - 4.0
Engine/transmission unit pivot bolt retaining nut	25 - 33	3.5 - 4.5

1 General description

The Honda Melody features an open scooter-type frame which consists of a large-diameter tubular forward section and a pressed steel rear section. The forward section of frame incorporates the steering head and footrest supports. Two metal support plates run parallel to the horizontal section of tube and are bolted to small tubular outrigger arms which in turn are welded to the main frame tube. The rear section of frame acts as a support for the engine/transmission unit, fuel and oil tanks and the seat. It consists of two sections which are spot welded together. Both the forward and the rear sections of frame incorporate numerous attachment points for the plastic body panels which cover the machine. The two sections of frame are joined together by a welded insertion joint.

The front forks of the machine comprise a tubular steel steering column which terminates in a plate steel bracket. This bracket acts to strengthen the joint between the steering column and the inverted U-section of tubular steel which is welded to its base. The legs of this U-section act as housings for the suspension components which provide the necessary insulation against shocks transmitted from the front wheel as it passes over irregular road surfaces. In order to provide the rigidity which is necessary to prevent the suspension components from distorting under load, a pressed steel arm is welded to the rear face of each fork leg. This arm is joined to the base of the fork rod and also to the spindle of the front wheel by a moving pivot arm, thus forming a leading link assembly. Neither fork leg is damped.

Rear suspension is by swinging arm, the arm in this instance being formed by the engine/transmission unit. An oil filled suspension unit serves to control movement of the arm and also provides the necessary damping action. This unit is non-adjustable. A mounting plate links the engine crankcase to its pivot point on the frame. The front of this plate contains two bonded-rubber bushes through which the pivot bolt passes. Bonded-rubber bushes are also incorporated in the crankcase to plate attachment and a rubber buffer prevents metal-to-metal contact between the upper surface of the plate and the crankcase. Attaching the engine/transmission unit to the frame by this method effectively absorbs any engine vibration which would otherwise be transmitted to the frame.

2 Front fork and steering head assembly: removal and refitting

1 The front fork and steering head assembly need only be removed as a unit from the machine if it is suspected that the steering head bearings are worn or if the front of the machine has sustained accident damage. Full information on removal and fitting of the fork components with the fork and steering head assembly still in situ in the machine are contained within the following Section of this Chapter.

2 Commence removal of the handlebar assembly from the steering stem by detaching both the handlebar nacelle and front body panel assembly from the machine. Full details for doing this are contained in Section 7 of this Chapter. Disconnect the headlamp unit by removing its bulb connectors from their locations in the reflector and place the unit in safe storage until required for reassembly.

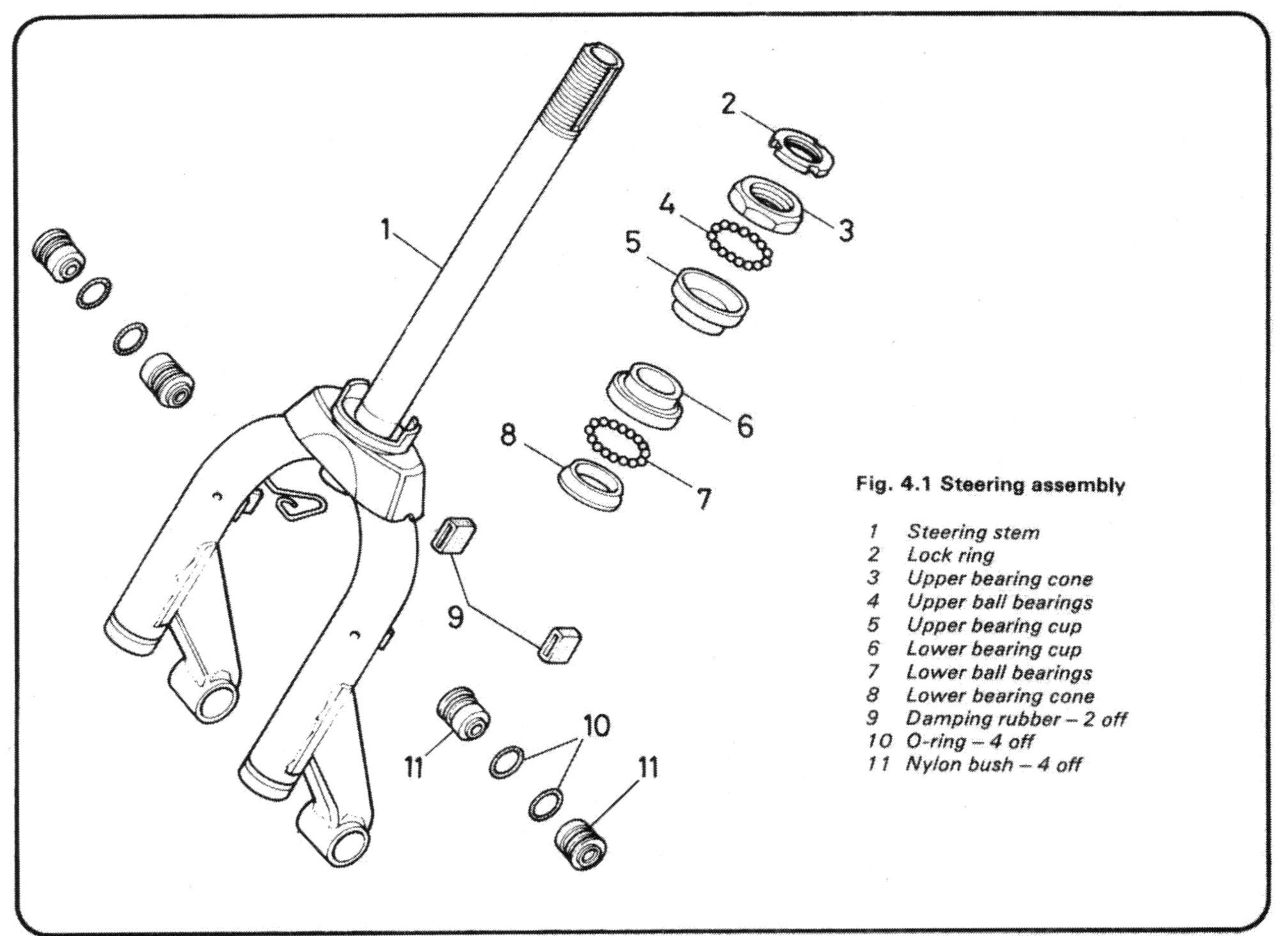

Fig. 4.1 Steering assembly

1 Steering stem
2 Lock ring
3 Upper bearing cone
4 Upper ball bearings
5 Upper bearing cup
6 Lower bearing cup
7 Lower ball bearings
8 Lower bearing cone
9 Damping rubber – 2 off
10 O-ring – 4 off
11 Nylon bush – 4 off

3 It is now necessary to disconnect any electrical wires and operating cables that will prevent the handlebars from being lifted clear of the top of the steering stem. Deciding which wire or cable to disconnect and how to disconnect it is really only a matter of common sense whilst noting the following points.

4 All electrical wires are colour coded, so there should be no confusion when it comes to reconnecting any disturbed connections. If in any doubt as to the fitted position of these wires, then make sure that they are clearly labelled before disturbing them. Note that the control cables which pass down through the steering stem must be fully withdrawn before removal of the steering head assembly can take place. Remember that it is only necessary to detach the wires and cables that will prevent the handlebars from being raised enough to clear the top of the steering stem; once the steering head assembly is freed, then the bars may be left to rest on top of the frame tube.

5 It is now necessary to raise the front of the machine far enough from the ground to allow not only removal of the front wheel but also to give enough clearance to enable the fork and steering head assembly to be lowered clear of the machine without its coming into contact with the ground. The best method of achieving this is to have an assistant lift the front of the machine whilst some substantial wooden blocks, or similar, are slid into position beneath the footrest panel mounting points. It is advisable that the assistant also helps to steady the machine during removal of the handlebar and steering stem retaining nuts.

6 Refer to Section 3 of Chapter 5 and remove the front wheel from the machine. With the handlebars held steady, remove the nut which retains them to the steering stem. Lift the handlebars off the steering stem and with the assistant supporting the fork legs to prevent them from dropping, unscrew the steering stem lockring, followed by the upper bearing cone. Remove the 26 steel balls contained in the upper bearing cup and then carefully lower the front fork and steering stem to clear the machine whilst taking care to retain the 26 steel balls that will fall from the lower bearing cup.

7 The various component parts of the steering head assembly should now be washed in clean petrol, whilst observing the necessary fire precautions, and given a careful visual examination when dry. Look for pits or scuff marks in the cup and cone bearing surfaces. If these are not smooth and polished in appearance, it will be necessary to renew them. The ball bearings should be renewed as a matter of course if the cups and cones have to be renewed. Other than this, they should be rejected if marked or damaged in any way. If the bearings are in anything other than perfect condition, the steering of the machine will be adversely affected, and for this reason any slightly suspect part demands renewal to ensure that the machine is kept roadworthy.

8 The upper and lower bearing cups may be removed from the headstock by passing a long drift through the inner bore of the headstock and drifting out the defective item from the opposite end. The drift must be moved progressively around the cup to ensure the item leaves the headstock evenly and squarely. The lower cone fits over the steering stem and may be removed by levering it upwards or by using a bearing extractor of suitable type. If difficulty is experienced with either of these operations, the assembly should be entrusted to an official Honda Service Agent who will have the necessary equipment to effect an economical repair. When fitting new items, ensure that they are located squarely in their fitted positions. Do not strike the bearing cups or cone directly with a hammer to drift them into position but support the component solidly and use a length of steel tube of the appropriate diameter as a drift in conjunction with the hammer. If the end of this tube is cut square to its length, then it will keep the cup or cone in question square in its location.

9 If it is found necessary to remove the front mudguard from its location over the fork legs, then this may be done by unscrewing the single bolt that retains it in position before lifting the mudguard over the top of the steering stem.

10 The steering head assembly is reassembled in the reverse order to that given for dismantling. Ensure that the bearings are generously lubricated with a high melting point, lithium based grease and that all 26 balls are included in each race. The bearings are adjusted for free play by slackening or tightening the upper bearing cone. It will be found that the cone can be tightened considerably from the finger-tight position without having any undue effect upon the ease with which the handlebars can be turned. This does mean, however, that a load of several tons is unwittingly applied to the bearings, which will be rapidly destroyed. When setting the cone, it is necessary to remove all discernible play, but no more. The cone is secured in position by tightening the lockring against it.

11 As a guide to the correct adjustment of the bearing cone, only very slight pressure should be needed to start the front wheel turning to either side under its own weight when it is raised clear of the ground. Check also that the bearings are not too slack; there should be no discernible movement of the forks, in the fore and aft direction.

12 When refitting the handlebars to the steering stem, align the tab projecting from the handlebar centre bracket with the groove cut in the steering stem before lowering the handlebars into position and fitting their retaining nut. This nut must be tightened to the specified torque loading of 43 - 65 lbf ft (6.0 - 9.0 kgf m).

13 Re-route and reconnect all disturbed electrical wires and control cables. Check that none of these wires or cables are likely to become chafed against moving cycle components and ensure that each disturbed control is in correct adjustment and operates smoothly and efficiently. Use this opportunity to lubricate the various controls. Full information on fitting the front wheel and adjusting both the front and rear brakes is contained in Chapter 5 of this Manual.

14 With the handlebar nacelle and the front body panel assembly refitted to the machine, fit the headlamp unit and align the beam in accordance with the instructions given in Section 24 of Chapter 6. Carry out a final check before taking the machine on the road to ensure that all the electrical switches mounted on the handlebar assembly function correctly and operate their respective components.

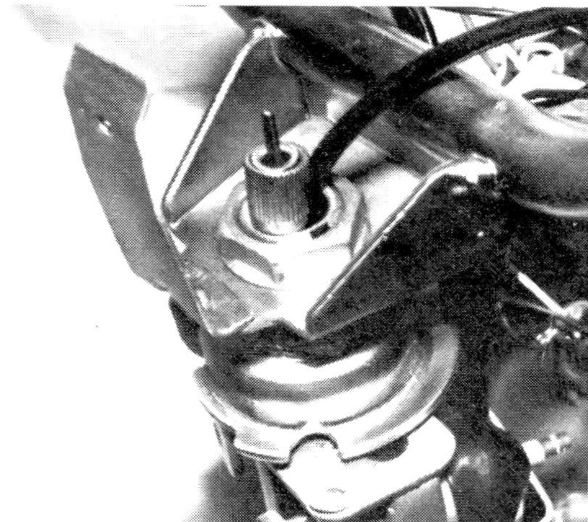

2.6 Remove the handlebar retaining nut

3 Front fork legs: dismantling, examination, renovation and reassembly

1 Commence removal of the components contained within each fork leg by placing the machine securely on its centre stand and placing a number of wooden blocks, or similar, beneath the footrest panel mounting points so as to prevent the

machine from tipping forward at any time during the following dismantling procedure. Remove the front wheel from the machine in accordance with the instructions given in Chapter 5 of this Manual.

2 It is advisable, before removing the component parts from each fork leg, to divide the work surface on which these parts are to be placed into two separate areas, one for each set of component parts. This will ensure that none of the parts become interchanged and, as a result, are fitted into the wrong fork leg. Dismantle each fork leg as follows.

3 Unscrew the nut and bolt which retains the fork pivot arm to the fork leg. Detach the pivot arm and push out each of the two nylon bush assemblies from the fork leg eye.

4 Detach the top of the fork gaiter from its location at the base of the fork leg tube and pull it down so that the fork rod retaining circlip is exposed. Remove this circlip and pull the complete fork rod assembly out of the fork leg tube. Dismantle this assembly and lay the component parts out on the prepared area of work surface after having carefully cleaned each one.

5 Examination of each set of fork leg components should begin with a check to ensure that the fork rod is not bent,. If the rod is bent or if its chromed surface has become badly corroded and pitted, then it must be renewed. Check that the rubber gaiter which protects the fork rod from the highly damaging effects of road dirt and salts is not split or perished. If this is found to be the case, then it must be replaced with a serviceable item.

6 Both the rubber buffer attached to the top of the fork rod and the rubber bush of the rod end eye should be inspected for signs of damage or deterioration and the necessary action taken if either component is found to be unserviceable. The rubber bush of the rod end eye, together with its metal centre, can be pushed out of the eye by using finger pressure.

7 Measure the length of the fork spring to check that it has not set to a length which is less than the given service limit of 112 mm (4.41 in). If the spring is too short, then it must be renewed. Note that it is also necessary to compare the spring of one fork leg with that of the other. If the springs are not of the same length, then an imbalance will occur between the fork legs which will result in the handling of the machine becoming adversely effected.

8 Check the condition of the spring seat for wear or damage and renew it if necessary. Note the rubber stop inserted into the fork spring and the stop onto which it abuts (this will most certainly have been retained in the fork leg tube). Inspect each component for damage or deterioration and renew as found necessary.

9 It is now necessary to inspect the tube of each fork leg for straightness. This can be done quite simply by laying a straight-edge along its length. Check that the surface of the tube is not dented. Any dent in the wall of the tube may well prevent the smooth passage of the fork rod and spring inside its length.

10 Inspect each pivot arm for damage and distortion. Check also that the holes drilled in its ends are not elongated and compare the two arms to check that the distance between hole centres is identical. Any defect in either arm will introduce an imbalance between the fork legs.

11 Each of the two nylon bush assemblies removed from the fork leg eyes will be seen to have O-rings fitted. If these rings have become flattened in use, or are split or perished, then they must be renewed. It is unlikely that the nylon bushes or their metal inserts will be worn to any unacceptable degree until a considerable mileage has been covered by the machine. Nevertheless, inspect the bearing surfaces of the bush assemblies for signs of scoring or wear and renew each assembly as required.

12 To complete the examination sequence, closely inspect the fork tube to pivot arm holding bracket weld for signs of fracture due to excessive forces being imposed upon it. Check also that the bracket is not distorted.

13 Reassembly of the fork leg component parts is a direct reversal of the dismantling procedure, whilst noting the following points. Check that the end of the spring with the smaller diameter is nearest the fork rod assembly and that the rubber stop inserted into the spring has its pointed end facing uppermost. Lubricate the inside of the fork tube and the fork rod assembly with a liberal amount of high melting point, lithium based grease before inserting the rod, complete with spring, into the fork tube. In practice, it was found that placing the end of a tommy bar against the underside of the rubber buffer of the fork rod and then pushing up on the end of the bar was by far the best method of forcing the buffer into the fork tube. It also helped if the buffer was inserted at an angle and its side faces well lubricated.

14 Check that the fork rod retaining circlip is properly located in its retaining groove before relocating the protective gaiter over the fork tube end. Lubricate the bearing surfaces of the nylon bush assemblies with grease before carefully pushing them into their location in the fork leg eye. Take care to ensure that the O-ring does not become detached from each bush. Fit the pivot arm to the fork leg and secure it in position by fitting its securing bolt and nut. Tighten the nut to the specified torque loading of 14 - 22 lbf ft (2.0 - 3.0 kgf m).

15 Finally, on completion of refitting the front wheel and unblocking the machine, place the machine on its wheels and push it forward against its front brake to check that the front fork legs are operating correctly.

3.4 Lay the fork leg component parts out for inspection

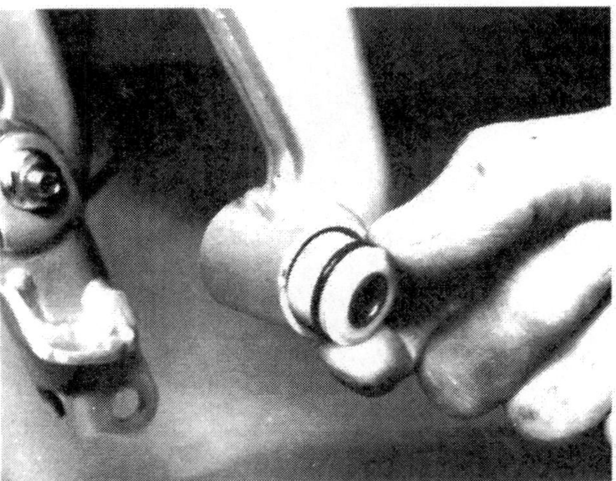

3.13a Locate each nylon bush in the fork leg eye ...

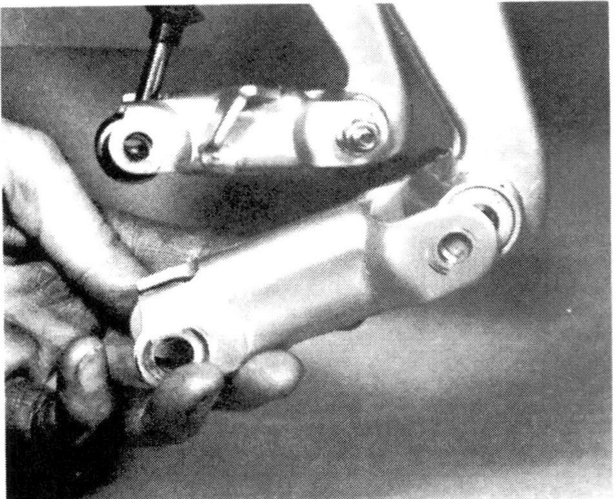
3.13b ... and fit the pivot arm

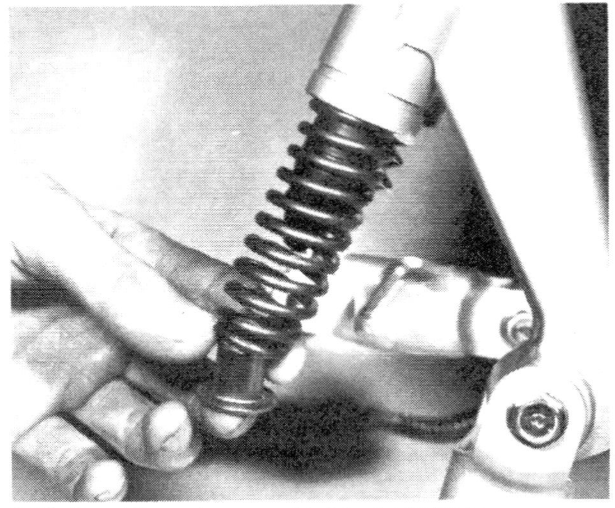
3.13c Insert the spring assembly into the fork tube ...

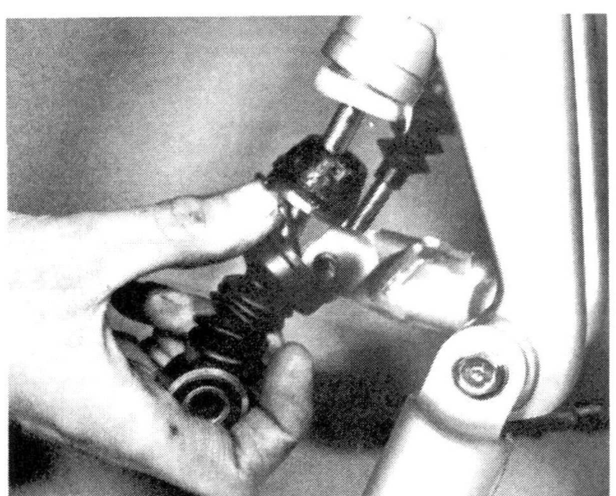
3.13d ... and follow with the fork rod assembly

3.14 Check that the retaining circlip is properly located

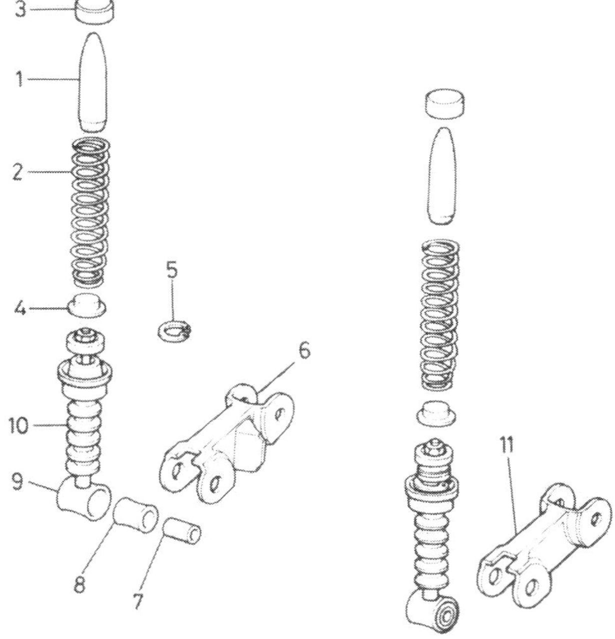

Fig. 4.2 Front forks

1 Rubber stop
2 Spring
3 Spring seat
4 Spring seat
5 Circlip
6 Right-hand pivot arm
7 Metal insert
8 Rubber bush
9 Fork rod
10 Gaiter
11 Left-hand pivot arm

4 Rear suspension pivot: examination and renovation

1 As already mentioned, rear suspension is provided by arranging the engine/transmission unit so that it is free to pivot around its front mounting point, the rear of the assembly being controlled by a single coil spring and shock absorber unit. A mounting plate serves to link the engine crankcase to a point on the frame. This plate is of pressed-steel construction, its forward end containing two bonded-rubber bushes. The plate is attached to the crankcase by a bolt which passes through the rear of the plate and through bonded-rubber bushes which are pressed into lugs cast into the crankcase. The plate is prevented from pivoting around this rear attachment by a centre mounting which takes the form of a plate which is slotted into the underside of the crankcase. Passing through this centre mounting is a stud which also passes through the unit mounting plate. The end of this stud is threaded to accept a retaining nut. Finally, in order to prevent metal-to-metal contact between the mounting plate and the crankcase, a rubber buffer is fitted between the two components.

2 Should wear have developed in the pivot assembly, it will be detected as side-to-side movement of the engine/transmission unit. This can be checked by grasping the rear of the transmission casing and pushing it from side to side whilst ensuring that the machine is held steady. Any discernible play will necessitate prompt attention before the resulting poor handling causes an accident.

3 Although it is possible to remove the mounting plate without the engine/transmission unit being removed completely from the machine, it was considered more satisfactory in practice to withdraw the unit completely so that proper attention could be paid to the bushes pressed into the crankcase lugs. Full instructions for unit removal are contained in Section 3 of Chapter 1.

4 Removal of the mounting plate from the crankcase, with the engine/transmission unit in or out of the frame, is a straightforward procedure. Commence by removing the nut from the centre mounting stud, together with the plate washer and rubber buffer fitted beneath it. If the unit has remained in the frame, position a support beneath the crankcase to prevent the unit from dropping and then unscrew the locknut from the end of the plate to frame attachment bolt. With the locknut and plain washer removed, use a soft-metal drift and hammer to drift the bolt from position.

5 Remove the nut and washer from the end of the plate to crankcase attachment bolt and drift the bolt from position. The mounting plate can now be detached and placed to one side, ready for examination. Pull the rubber buffer off the centre mounting stud and detach the centre mounting plate from its retaining slots in the crankcase.

6 Carry out a close examination of the bonded-rubber bushes contained in both the mounting plate and the crankcase lugs. Each bush is made up of a rubber cylinder which is bonded to an external and an internal steel sleeve. Inspect each bush for deterioration of the rubber, wear of the inner sleeve and for separation between the rubber and either sleeve. If any such fault is evident then bush renewal is required.

7 The bushes fitted to the crankcase are a tight drive fit in their housing lugs. When the time comes for renewal, it will probably be found that, due to corrosion between the dissimilar metals (aluminium and steel), the already tight bushes have become almost immovable. As a means of removal, attempting to drive the bushes out will probably prove unsuccessful, because the rubber will effectively damp out the driving force, and damage to the lugs may occur. It is suggested that the bushes are drawn from position using a fabricated puller as shown in the accompanying diagram. This can be made from a short length of thick-walled tube, the inside diameter of which is slightly larger than the outside diameter of the bush, and two thick plate washers, one of which has an outer diameter slightly smaller than that of the bush outer sleeve and the other having a diameter which is greater than that of the tube. It is advisable to use a high tensile bolt and nut, which will be better able to take the stresses involved.

8 Heating the crankcase lugs with boiling water to expand the alloy and introducing penetrating oil around the bush to lug joint are both methods of helping to free the bushes before applying force to them by using the fabricated puller. If, having used every means possible, it is found that the bushes are reluctant to move, it is recommended that the engine/transmission unit be returned to a Honda service agent whose expertise can be brought to bear on the problem. Do not attempt any method of bush removal that may crack or damage the crankcase lugs.

9 Removal of the bushes fitted in the mounting plate can be achieved by supporting the plate on wooden blocks so that one of the bushes can be drifted downwards and out of its location. To do this, insert a long metal drift through the centre of the opposite bush so that its end abuts against the edge of the outer sleeve of the bush to be removed. Work the drift around the bush sleeve so that the bush remains square to the mounting plate during removal. A fairly heavy hammer will be needed to strike the drift in order to effect the initial freeing of the bush, Again, the initial use of heat and penetrating oil will aid removal of the bush. It must be noted that the bushes fitted in the mounting plate are not listed as separate items to the plate itself. Rather than incur the expense of buying a complete mounting plate assembly, it is well worth contacting a motor factor or any of the suppliers of bearings and bushes who advertise in the Yellow Pages or motoring trade magazines as to whether a similar type of bush can be supplied.

10 All new bushes may be driven into their locations by using a tubular drift against the outer sleeve. If this method is used, make sure that the opposite end of the bush location is well supported. In the case of the crankcase bushes, reverse the operation of the puller to draw each bush into position. Take care to check that both the bush outer surface and the bore of the bush location are free from all traces of corrosion and are cleaned of all contamination. Smearing a thin film of grease over the surface of the location bore will aid insertion of the bush. Take care to remove all excess grease from around the bush after insertion.

11 Continue the examination sequence by inspecting the mounting plate, the crankcase lugs and the frame mounting points for signs of fatigue or failure. Any fatigue in the crankcase lugs may well be indicated by a series of very fine hairline cracks. It is unlikely that such cracking will be seen unless the lugs have been thoroughly cleaned and degreased. Any ovality of the holes in the frame mounting points will necessitate specialist work by a skilled welder and machinist, who will have to fill the holes with metal before redrilling them to accept an unworn pivot bolt.

12 Inspect the shafts of both attachment bolts for signs of wear. This is not likely to be found on the crankcase to mounting plate bolt but may well be seen in the form of two grooves where the frame to mounting plate bolt forms a contact with the frame mounting points.

13 The centre mounting components should be inspected for general wear, damage and deterioration, and renewed as found necessary. The centre mounting plate retaining slots cut in the crankcase should be inspected for signs of fatigue or failure.

14 With the pivot assembly thus examined and renovated, proceed to refit it as follows. Slot the centre mounting plate into the crankcase and push the serviceable rubber buffer into position over the centre mounting stud. Place the unit mounting plate over the centre mounting stud and fit the second rubber buffer followed by the plate washer and retaining nut. Do not tighten this nut.

15 Align the unit mounting plate with the centre of each crankcase mounting and carefully drift the rear mounting bolt into position. Applying a light coating of grease to the shank of this bolt will ensure that it does not become corroded to the centre of the bonded rubber bushes, thus making extraction of the bolt at a later stage relatively trouble free.

16 Tighten the centre mounting nut to a torque loading of 14

– 22 lbf ft (2.0 – 3.0 kgf m). Fit the nut and washer to the rear mounting bolt and tighten the nut to a torque loading of 25 – 33 lbf ft (3.5 – 4.5 kgf m). Refer to Section 43 of Chapter 1 for full details of refitting the engine/transmission unit into the frame. If the unit was not removed from the machine, note that the specified torque loading for the mounting plate to frame attachment bolt securing nut is 25 – 33 lbf ft (3.5 – 4.5 kgf m).

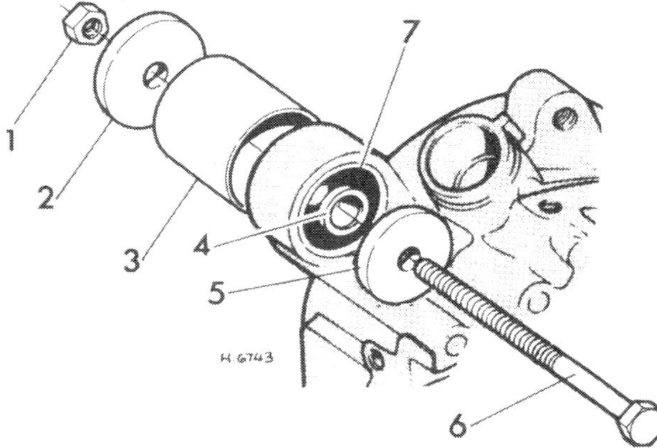

Fig. 4.3 Crankcase bush removal tool

1 High tensile nut
2 Thick washer
3 Pipe
4 Bush inner sleeve
5 Thick washer
6 High tensile bolt
7 Bush outer sleeve

4.4 Remove the nut from the centre mounting stud

4.5a Remove the crankcase attachment bolt and detach the plate ...

4.5b ... to expose the rubber buffer and centre mounting stud

5 Rear suspension unit: removal, examination, renovation and refitting

1 The rear suspension unit fitted to the Honda Melody can be removed from the machine simply by unscrewing the bolt which retains it to the transmission casing and then unscrewing the single flange nut which retains it to the frame mounting. This operation must be carried out with the machine supported properly on its centre stand so that the transmission casing is allowed to drop away from the suspension unit once the retaining bolt is removed. Pull the lower mounting of the unit rearwards to clear the transmission casing and then pull the unit sideways clear of the frame.

2 The suspension unit comprises a hydraulic damper (effective primarily on rebound), a concentric spring and a rubber stop. The unit is secured to the frame and transmission casing through rubber-bushed lugs. The top lug of the unit contains a rubber insert which can be displaced fairly easily by the application of finger pressure. The lower bush is a press fit into the transmission casing and is more difficult to remove; the procedure for doing so being described in Section 4 of this Chapter (paragraphs 7 and 8).

3 Carry out a careful examination of the suspension unit whilst noting the following points. Inspect the damper unit for signs of leakage, corrosion or damage to the chromed surface of the piston shaft, deterioration of the seal, rubber buffer and rubber mounting bushes, and damage to the piston housing. The damper unit can only be checked for efficiency after the suspension unit has been dismantled. There is no means of draining or topping up the fluid in the unit because it is sealed during manufacture. If signs of fluid leakage are apparent, then the complete damper assembly must be renewed.

4 Check also for straightness of the damper rod. The damper unit must be renewed if this rod is seen to be bent. Breakage of the concentric spring will be obvious and renewal of the spring will necessitate its detachment from the damper unit.

5 The suspension unit can be dismantled by compressing the spring against the top of the damper unit and then removing the lower mounting attachment from the threaded end of the damper rod. Honda supply a special tool for doing this job. This tool should be used in conjunction with the two special adaptors also supplied by Honda (Nos 07967-GA70101 and 07967-GA70200). If none of these tools can be borrowed or hired from an official Honda service agent, or the unit cannot be returned to the agent for renewal of the defective component, then it is recommended that the following procedure be used.
6 Invert the unit and place it upright on a strong work surface. If possible, clamp the lug of the damper unit between the protected jaws of a vice. With an assistant pulling down on the end of the spring so that it clears the upper facing mounting attachment, insert an open-ended spanner over the locknut located beneath the mounting attachment and proceed to unscrew the attachment by using a spanner fitted over its sides or a tommy bar passed through its bolt holes. Note that it is essential an assistant keeps the spring pressure from acting on the mounting attachment during this operation, because once the attachment reaches the end of its thread, spring pressure will cause it to fly off the damper rod whilst taking the end threads with it. Both the spring and rubber stop can now be drawn off the damper unit.
7 With the suspension unit thus dismantled, temporarily refit the mounting attachment to the damper rod and compress and extend the damper. If the unit is serviceable, little resistance will be felt when compressing the damper, whereas heavy resistance should be felt as the unit is extended.
8 The unit spring must be renewed if it has become fatigued or has set to a length which is less than the service limit of 175.5 mm (6.91 in).
9 Reassembly and fitting of the unit is a direct reversal of the removal and dismantling procedures, whilst noting the following points. Ensure that the unit has been thoroughly cleaned before reassembly, especially around the area of the chromed piston rod and its seal. Do not grease the piston rod.
10 Fit the spring with its close coils adjacent to the lug of the damper unit. Apply a locking compound to the threads of the damper rod before fitting the locknut and mounting attachment. Tighten the locknut to a torque loading of 11 – 14 lbf ft (1.5 – 2.0 kgf m).
11 With the unit fitted to the machine, tighten the retaining nut to a torque loading of 22 – 29 lbf ft (3.0 – 4.0 kgf m) and the bolt to a torque loading of 18 – 25 lbf ft (2.5 – 3.5 kgf m). Move the machine off its centre stand and push down on the rear of the machine several times to ensure that the suspension unit is functioning correctly.

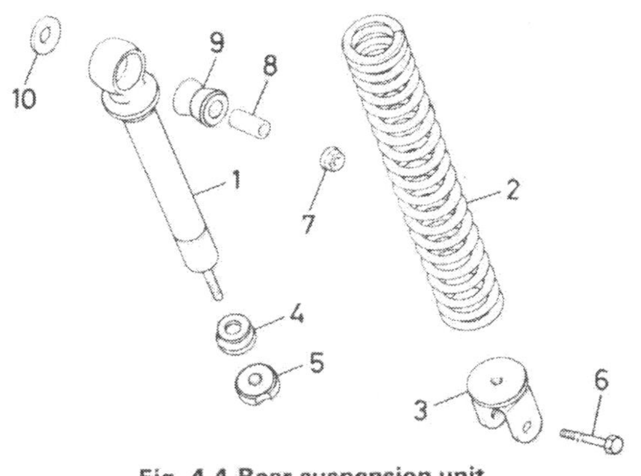

Fig. 4.4 Rear suspension unit

1 Hydraulic damper
2 Spring
3 Lower mounting eye
4 Rubber stop
5 Nut
6 Bolt
7 Nut
8 Metal insert
9 Rubber bush
10 Washer

5.1a Remove the suspension unit to transmission casing retaining bolt ...

5.1b ... before detaching the unit from its frame mounting

5.2 Inspect the suspension unit lower mounting bush

6 Frame: examination and renovation

1 If the machine is stripped for a complete overhaul, this affords a good opportunity to inspect the frame for cracks or other damage which may have occurred in service. Check the points at which the lower section of frame tube joins both the steering head and the pressed steel rear section of frame; these are the points where fractures are most likely to occur. Checking alignment of the steering head tube with the vertical section of the pressed steel section of frame will show whether the machine has been involved in a previous accident.
2 Check carefully areas where corrosion has occurred on the frame. Corrosion can cause a reduction in the material thickness and should be removed by use of a wire brush and derusting agents. After the machine has covered a considerable mileage, it is advisable to examine the frame closely for signs of cracking or splitting at the welded joints.
3 If the frame is broken or bent, professional attention is required. Repairs of this nature should be entrusted to a competent repair specialist, who will have available all the necessary jigs and mandrels to preserve correct alignment. Repair work of this nature can prove expensive and it is always worthwhile checking whether a good replacement frame of identical type can be obtained at a reasonable cost.
4 Remember that a frame which is in any way damaged or out of alignment will cause, at the very least, handling problems. Complete failure of a main frame component could well lead to a serious accident.

7 Body panels: removal and refitting

1 It will be seen that most of the Honda Melody is covered by removable panels, thereby necessitating a certain amount of preliminary dismantling in order to gain access to any one component part located beneath them. The procedure for removing and fitting any one of these panels is given in the following paragraphs of this Section. Bear in mind that these panels are constructed of a soft plastic and will therefore become damaged if not treated with some degree of care. Directly it is removed, place each panel on a soft protected surface, well away from where it may be kicked or hit by flying tools.

Handlebar nacelle
2 To detach the upper and lower halves of the handlebar nacelle, start by removing the single screw from the centre of the rear face of the lower half of the handlebar nacelle. Follow this by removing the two screws which pass upwards through the upper edge of the lower half of the nacelle and then manoeuvre the nacelle down and rearwards to clear the handlebar assembly.
3 Detach the speedometer cable from the base of the instrument head by unscrewing its knurled retaining cap and then pulling the cable downwards to clear its drive attachment. Remove each of the two headlamp unit retaining bolts from the upper half of the handlebar nacelle and carefully pull the headlamp unit from position. Lift the upper half of the nacelle up

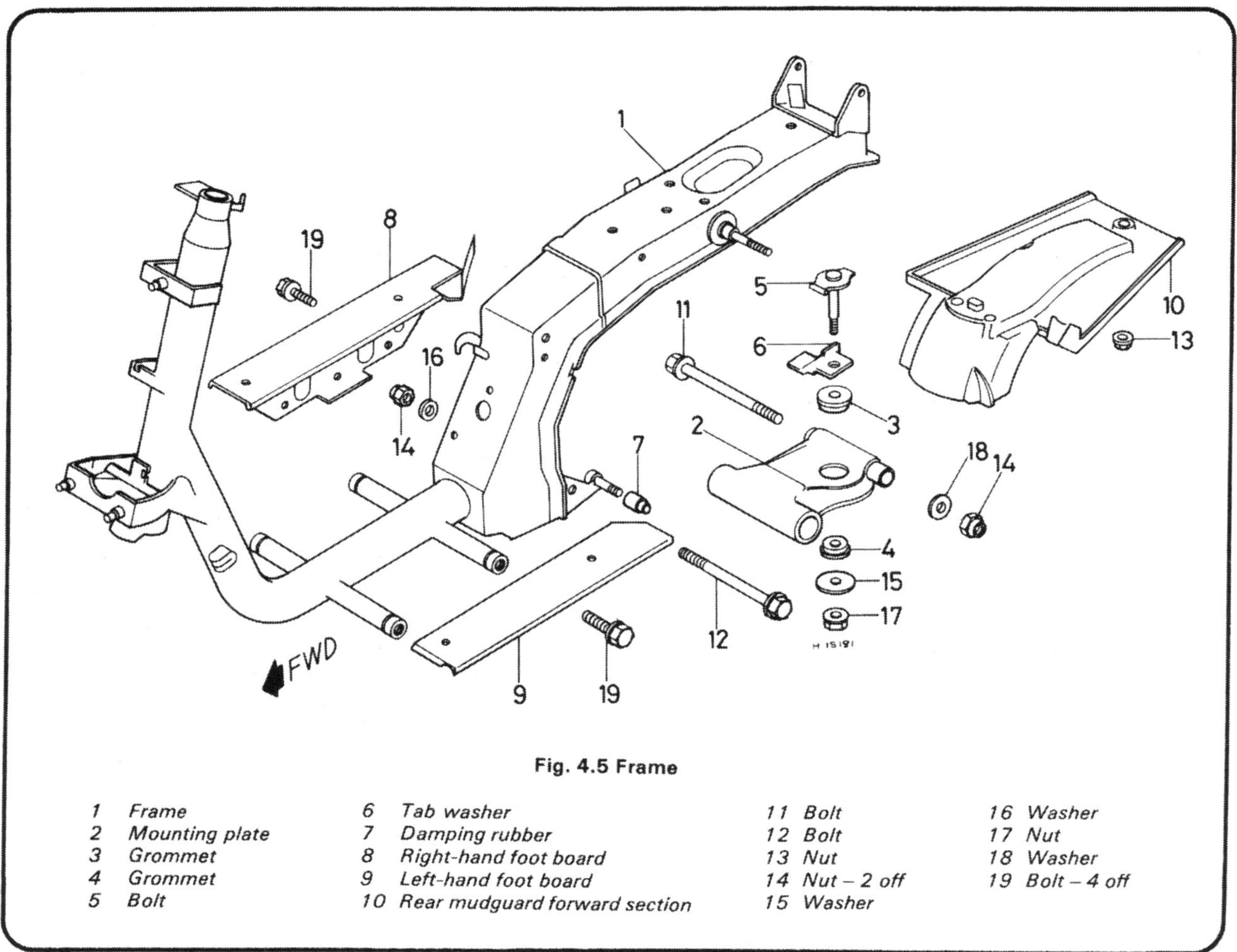

Fig. 4.5 Frame

1 Frame
2 Mounting plate
3 Grommet
4 Grommet
5 Bolt
6 Tab washer
7 Damping rubber
8 Right-hand foot board
9 Left-hand foot board
10 Rear mudguard forward section
11 Bolt
12 Bolt
13 Nut
14 Nut – 2 off
15 Washer
16 Washer
17 Nut
18 Washer
19 Bolt – 4 off

away from the handlebars and unplug each bulb holder from the instrument head.

4 Fitting of the nacelle assembly to the handlebars can be achieved by reversing the removal procedure. Before taking the machine onto the road, refer to Section 24 of Chapter 6 and check the alignment of the headlamp beam.

Front body panel

5 Removal of the two halves of the front body panel will also necessitate removal of the carrier pocket and the luggage basket. Note that only the Melody deluxe model (the NS 50MS) has the carrier pocket fitted as standard. To detach this pocket, simply release its two mounting clips by loosening their retaining screws.

6 The luggage basket is held in position by the three dome headed nuts which also retain the front half of the body panel to its frame mountings. It was found in practice, that a thin-walled socket was the only form of tool that would provide an easy means of loosening these three nuts. With the basket pulled clear of the machine, move the top centre of the rear half of the body panel and remove the single screw which serves to retain it in position.

7 Move to each side of the front half of the body panel where it is secured to the footrest panel by a single nut and bolt. Remove each of these nuts and bolts and detach both halves of the front body panel from the machine. Fitting of the panel halves is a direct reversal of the removal procedure.

Footrest panel

8 Before attempting to move the footrest panel from its frame mountings, it will first be necessary to detach the centre body panel so that the rear of the footrest panel can be lifted. To further ease removal of the footrest panel, it may also be necessary to detach the complete front panel assembly. In practice, it was found that the footrest panel could be removed without detachment of the front panel but some amount of distorting of the panel was necessary. In either case, removal of the two nuts and bolts which secure the footrest panel to the front panel is necessary before unscrewing the four footrest panel retaining bolts and then lifting the panel clear of its mounting points, rear end first. Fitting of the footrest panel is a direct reversal of the removal procedure.

Centre body panel

9 The centre body panel is retained in position by one dome head nut with washer and a single screw. With the nut and screw removed, the panel can be detached from the machine simply by easing it forward clear of its mounting points. Fitting of the panel can be achieved by reversing this removal procedure.

Rear body panel

10 Commence removal of the rear body panel by unscrewing the single flange nut which retains the rear mudguard to the underside of the rear body panel. This nut is located in the centre of the rear edge of the mudguard. With the nut removed, the mudguard can be eased down and rearwards to free it from its forward locating tabs.

11 Raise the seat of the machine and remove the battery from its housing by unscrewing the clamp retaining screw, lifting the clamp, removing the battery cover, disconnecting the battery leads and then carefully lifting the battery from position whilst noting the location of its vent pipe in the housing. Remove the rubber pad from the base of the battery housing to reveal the bolt located beneath it. Unscrew this bolt and remove the battery holder.

12 Remove the luggage rack from its mounting points by unscrewing its two retaining bolts and two dome nuts with washers. Remove the two bolts which are located directly beneath the tail/stop lamp lens. Ease the lamp unit, complete with number plate, downwards to release it from its upper retaining clip. Pull the lamp unit away from the machine whilst carefully pulling the electrical wires out of their location in the body panel. Once the push connectors for these wires are exposed, note the colour coding of each wire and then pull apart the connectors to allow the lamp unit to be lifted away from the machine and placed in safe storage ready for refitting.

13 Remove the mounting bracket for the tail/stop lamp unit by unscrewing its two retaining bolts. Remove the oil tank filler cap and the small drip tray located beneath it. Carefully ease the rear body panel upwards over the oil tank filler point and then detach the battery vent pipe from the retaining stub on the underside of the battery housing. The rear body panel can now be lifted from position. Refit the oil tank filler cap directly the panel has been removed so as to obviate any chance of contamination entering the tank.

14 Fitting the rear body panel and its associated components is a direct reversal of the removal procedure. Remember to check the operation of the tail and stop lamps before taking the machine onto the road.

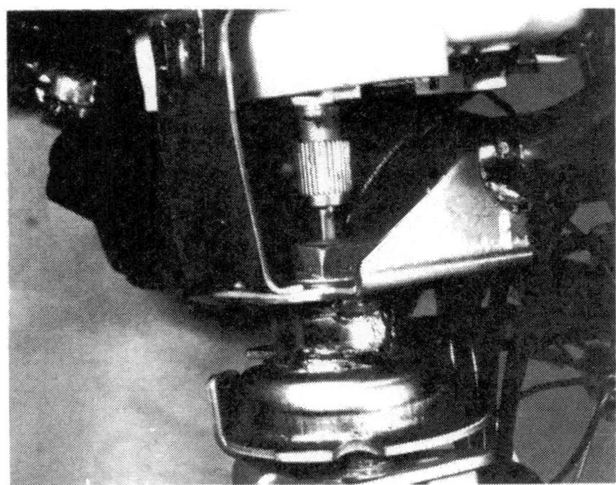

7.3 Detach the speedometer cable from the instrument head

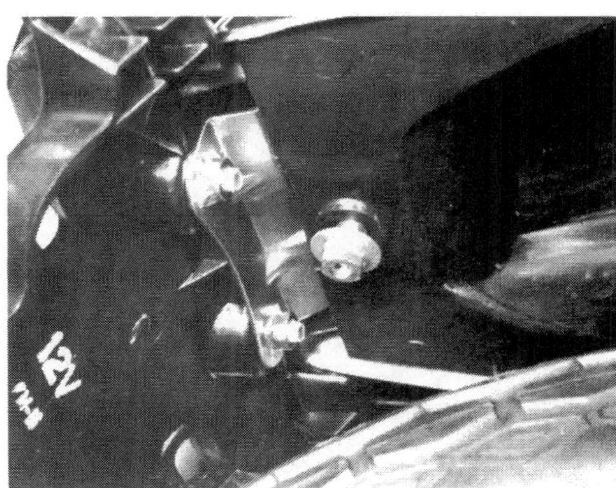

7.10 Remove the rear mudguard retaining nut

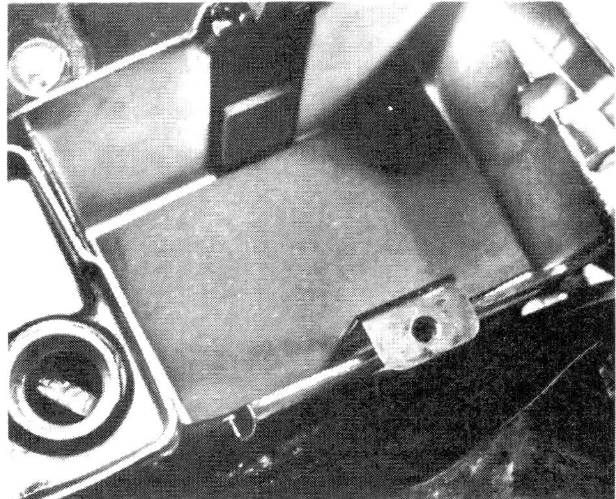
7.11a Remove the rubber pad from the battery housing ...

7.11b ... to reveal the battery holder retaining bolt

7.12a Remove the tail lamp unit retaining bolts

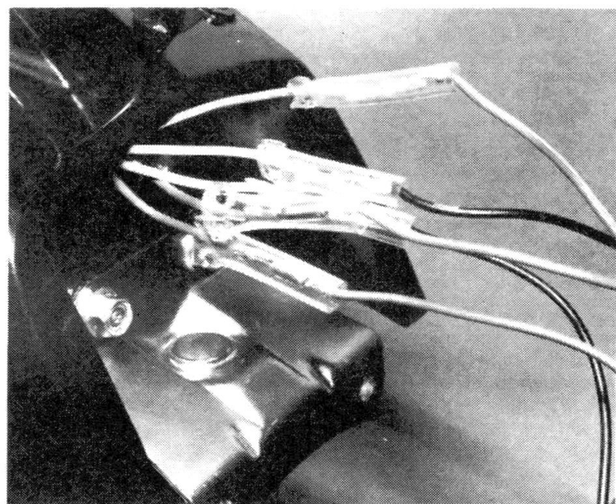
7.12b Expose the tail lamp unit wiring connectors

7.13a Remove the tail lamp unit mounting bracket

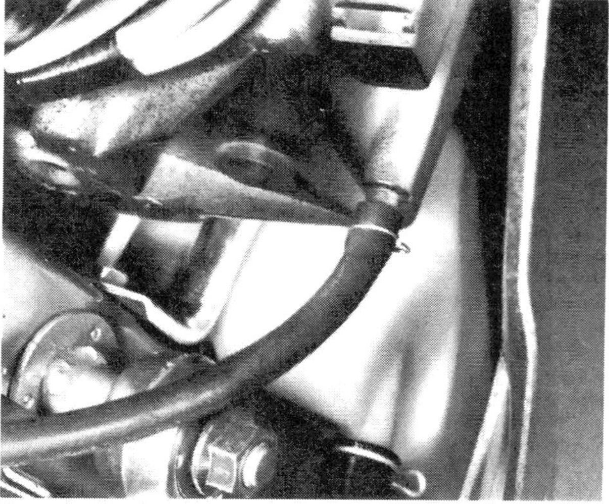
7.13b Detach the battery vent pipe from its retaining stub

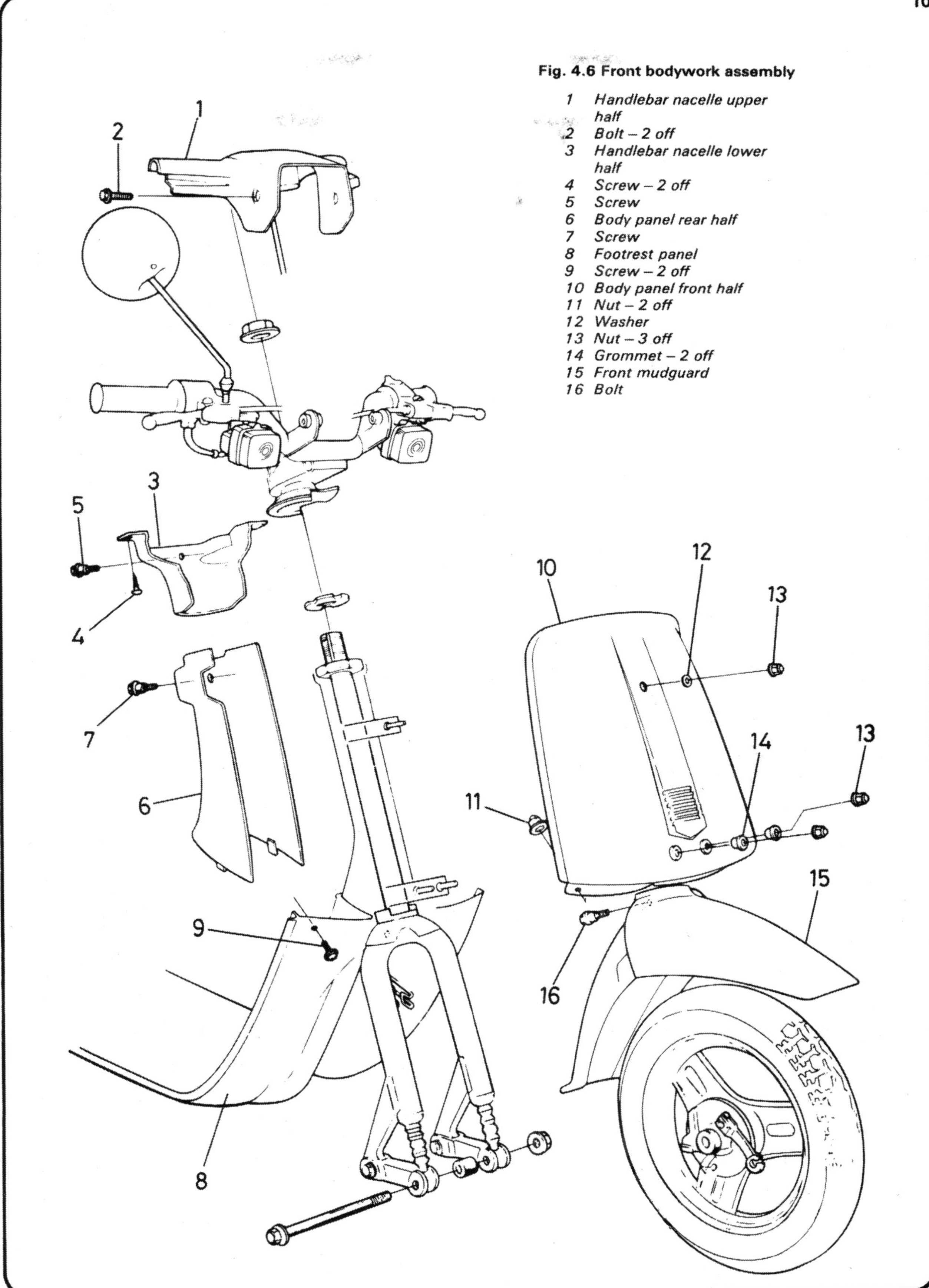

Fig. 4.6 Front bodywork assembly

1 *Handlebar nacelle upper half*
2 *Bolt – 2 off*
3 *Handlebar nacelle lower half*
4 *Screw – 2 off*
5 *Screw*
6 *Body panel rear half*
7 *Screw*
8 *Footrest panel*
9 *Screw – 2 off*
10 *Body panel front half*
11 *Nut – 2 off*
12 *Washer*
13 *Nut – 3 off*
14 *Grommet – 2 off*
15 *Front mudguard*
16 *Bolt*

8 Seat: removal and refitting

1 The seat of the Honda Melody is mounted atop the fuel tank. It is retained in position by a hinge which is secured to the front section of the seatpan and to the front of the fuel tank. Two rubber buffers prevent the seatpan from coming into direct contact with the rear of the fuel tank when the seat is lowered and a seat lock is provided to secure the seat in its down position. Note that the seat is designed to take the weight of no more than one person.
2 To remove the seat, detach its lock ring from the seat lock and raise the seat. Unscrew the two flange nuts which retain the seat hinge to the fuel tank and then lift the seat clear of the machine. Fitting the seat is a direct reversal of the removal procedure. Check that the seat is correctly aligned on the fuel tank before finally tightening the two securing nuts and always lock the seat in its down position.

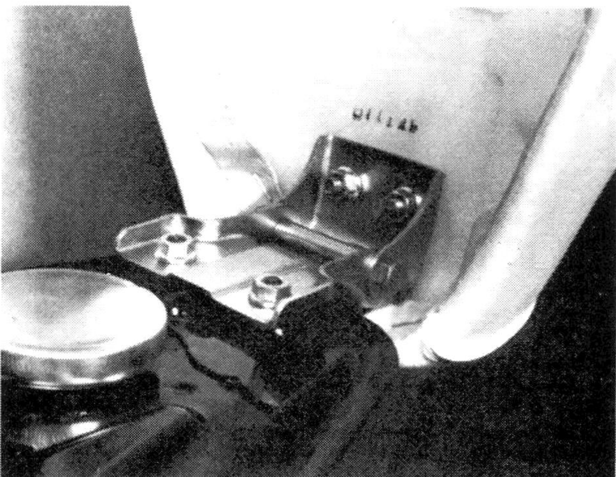
8.2 Remove the seat hinge retaining nuts

9 Steering lock: location and renewal

1 The steering lock and ignition switch are combined in one unit which is mounted on a plate which forms part of the steering head assembly. When the key is turned to its 'Lock' position, a bar projects from the lock body and through a cutout in the lower periphery of the handlebar centre bracket, thus locking the handlebars in one set position.
2 In order to gain full access to the lock, it is necessary to remove the carrier pocket from the rear of the front body panel, followed by the rear section of the front body panel. Note that only the Melody deluxe model (the NS 50 MS) has the carrier pocket fitted as standard and that full instructions for the removal and fitting of body panels and cycle components may be found in an earlier Section of this Chapter. The lock is secured to its mounting plate by the two countersunk cross head screws. Removal of the innermost of these two screws will necessitate removal of the handlebar assembly.

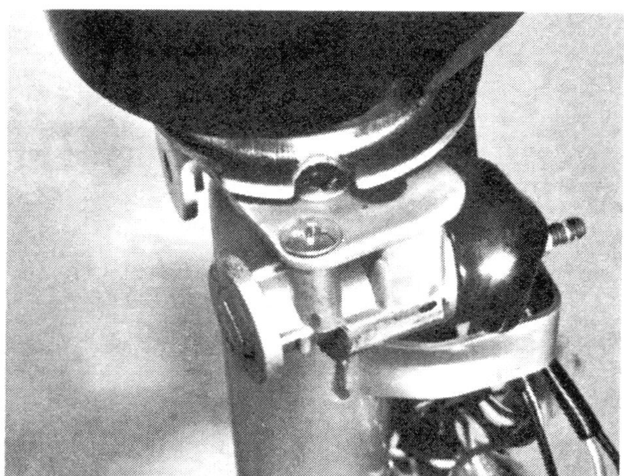
9.1 The steering lock is retained in position by two cross head screws

10 Centre stand: examination and maintenance

1 The Honda Melody is equipped with a centre stand which is formed out of tubular steel. The stand pivots on a steel shaft which passes through two plates on the frame structure and is retained in position by a plain washer and a split-pin.
2 Maintenance of the stand consists of checking that it retracts smoothly and fully against its stop. The stand is caused to retract by a spring which is connected between a point on the stand and a point on the frame structure. Closely examine this spring for any signs of fatigue or wear at the points where it passes through the stand or the frame. If the stand is slow to return against its stop, then it is likely that its pivot points are in need of lubrication and probably need cleaning of any hardened deposits or corrosion.
3 Should the stand require any attention, then it can be removed by removing the split-pin and the plain washer and then displacing the pivot shaft. Once this is done, the spring can be unhooked from the stand and the stand manoeuvred clear of the machine. Check for wear at the bearing surfaces of the shaft to stand. Excessive wear at these points will affect the efficiency of the stand in retracting. The same applies if excessive wear is seen at the spring retaining holes.
4 Note that fitting of the stand can prove rather arduous because it is necessary to manoeuvre the stand into position against spring pressure whilst the pivot pin is fitted. There is no easy way of accomplishing this operation, and a mixture of patience and strength is essential. It is almost indispensable to have some assistance during reassembly. Remember to fit a new split-pin and to retain it in position by correctly bending apart its legs.

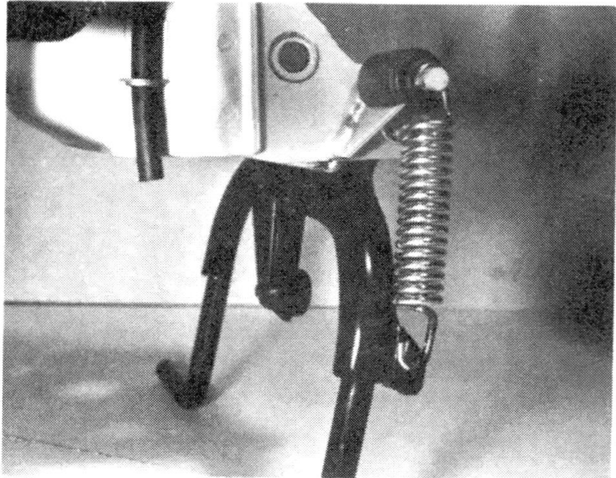
10.2 Closely examine the centre stand spring

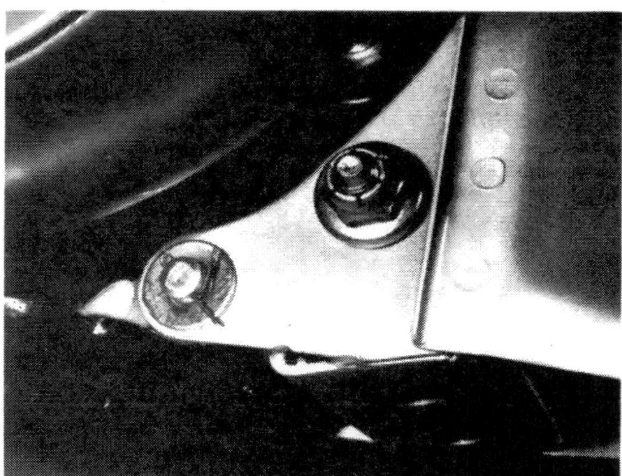

10.4 Do not omit to fit the pivot pin retaining washer and split-pin

11 Kickstart lever: examination and renovation

1 The kickstart lever is secured to its splined shaft by means of a pinch bolt which passes through the edge of its mounting boss. The inner surface of the lever boss is splined to correspond with the shaft. The footrest section of the lever is designed to swivel about the end of the main section of lever, so that when the engine is started, it can be tucked out of harm's way against the transmission casing.
2 A spring-loaded ball bearing locates the footrest section of the lever in either the operating or folded position. If the action of this mechanism becomes sloppy it is probable that the spring behind the ball bearing needs renewing. Unfortunately, Honda do not design the lever so that this mechanism can be dismantled; it is therefore necessary either to purchase a new item or enquire at a 'bike breaker' as to the availability of a second-hand serviceable item.
3 If the action of the folding mechanism is stiff, then apply a penetrating fluid to free it before lubricating it with clean motor oil. Do not ride the machine with the footrest section of the lever in the operating position as it is possible that in the unfortunate occurrence of the machine being 'dropped', the end of the footrest section striking the ground will serve to push the main section of lever into the transmission casing with the resulting breakage and expense.
4 If the kickstart lever is bent in an accident, then it should be removed and straightened by using the following procedure. Unscrew and remove the pinch bolt from the lever boss and pull the lever off its shaft. Clamp the lever in between the jaws of a vice and heat the area around the deformed section to a dull red heat by using the flame of a blowlamp. Straighten the lever and allow it to cool naturally by means of the surrounding airflow. Cooling the lever by immersing it in cold water is not only a dangerous practice but will cause the affected section of shaft to become embrittled.
5 If there is evidence of failure in the metal of the lever either before or after straightening, then it is essential that the damaged component be renewed. If the kickstart lever breaks whilst being used, then serious injury could result to the leg or foot. When fitting the lever, check that the alignment dot on the end of its shaft is adjacent to the gap in the lever boss and ensure that the lever is not in contact with the transmission casing at any point along its length.

12 Speedometer head: removal and refitting

1 The instrument head itself is generally reliable, and is the least likely culprit in the event of failure; this normally being attributable to the cable rather than the instrument mechanism. If, however, it is noted that the speedometer has ceased to function whilst the odometer (mileage recorder) still functions, the instrument can be assumed to have failed. No form of repair is practicable at home, and a replacement speedometer will be required. The only alternative is to seek the assistance of one of the companies who specialise in this type of repair work.
2 To gain access to the instrument head, start by removing the single screw from the centre of the rear face of the lower half of the handlebar nacelle. Follow this by removing the two screws which pass upwards through the upper edge of the lower half of the nacelle and then manoeuvre the nacelle down and rearwards to clear the handlebar assembly.
3 Detach the speedometer cable from the base of the instrument head by unscrewing its knurled retaining cap and then pulling the cable downwards to clear its drive attachment. Remove each of the two headlamp unit retaining bolts from the upper half of the handlebar nacelle and carefully pull the headlamp unit from position. Lift the upper half of the nacelle up away from the handlebars and unplug each bulb holder from the instrument head.
4 It is now possible to detach the instrument head from the upper half of the handlebar nacelle by removing the two self-tapping screws which secure it in position. Before doing this, note that if the instrument head is turned upside down at any point during the dismantling sequence, then the lens of the instrument may well become contaminated by any lubricant contained within the unit.
5 It will be seen that the speedometer head is contained within a plastic casing which also provides a housing for the fuel level gauge, the oil level warning light and the direction indicator warning light. This casing can be split into two halves to allow for removal of the speedometer head and fuel level gauge by unscrewing the four screws that pass through the instrument lens and into the casing body. Take care whilst separating the lens from the casing body not to tear the sealing gasket located between the two components; any break in the seal formed by this gasket will allow moisture to enter the components contained within the casing.
6 Reassembly of the instrument casing and refitting of the serviceable component to the machine can be achieved by reversing the removal and dismantling procedures. Before taking the machine onto the road, refer to Section 24 of Chapter 6 and check the alignment of the headlamp beam.

13 Speedometer drive and cable: examination

1 Drive from the speedometer drive gearbox is transmitted to the instrument head by way of a flexible cable. This flexible cable consists of a resilient but torsionally rigid inner cable which runs inside, and is protected by, a reinforced outer cable. This arrangement allows the drive to pass through gentle bends and absorbs the relative movement between the front wheel and the instrument.
2 Although considered to be flexible, it is preferable to ensure that the cable does not pass through acute bends, which would shorten its effective life. The straighter the cable's run, the lower the rate of wear and risk of breakage.
3 If the speedometer ceases to function, suspect a broken cable. Inspection will show whether the inner cable has broken.
4 Spin the inner cable to check for resistance. Most cables have a tight spot, but if the resistance is severe and a wavering speedometer has been noted, the cable should be renewed. Lubrication is difficult with this type of cable, but an aerosol chain grease or a silicone-based lubricant can often be introduced using the aerosol's thin extension nozzle.
5 Drive to the cable takes the form of a gear mechanism built into the front brake backplate. This is not normally prone to wear, and maintenance can be restricted to inspection and regreasing whenever the front brake components receive attention. Any mechanical failure will be obvious and will require renewal of the damaged parts.

14 Fault diagnosis: frame and forks

Symptom	Cause	Remedy
Machine veers to left or right with hands off handlebars	Incorrect wheel alignment	Rectify cause of misalignment
	Bent front forks	Renew
	Twisted frame	Renew
Machine rolls at low speeds	Overtight steering head bearings	Slacken and retighten upper bearing cone to specified amount
Machines judders upon application of front brake	Slack steering head bearings	Tighten upper bearing cone to specified amount
	Distorted front brake drum	Refer to Chapter 5
Machine pitches badly on uneven surfaces	Ineffective front forks	Service or renew as necessary
	Defective rear suspension unit	Renew
Fork action stiff	Fork components dry or dirty	Dismantle, clean and lubricate
Machine wanders. Steering imprecise, rear wheel tends to hop	Worn rear suspension pivot assembly	Dismantle, examine and renovate

Chapter 5 Wheels, brakes and tyres

Refer to Chapter 7 for information on the NB, ND and NP50 models

Contents

Specifications

Wheels		
Type	Pressed steel	
Rim runout service limit:		
Axial	2.0 mm (0.08 in)	
Radial	2.0 mm (0.08 in)	
Front wheel spindle runout service limit	0.2 mm (0.008 in)	
Brakes	**Front**	**Rear**
Type	Single leading shoe, drum	
Shoe lining thickness	–	5.0 mm (0.20 in)
Service limit	2.0 mm (0.08 in)	2.0 mm (0.08 in)
Drum ID	–	95 mm (3.74 in)
Service limit	80.5 mm (3.17 in)	95.5 mm (3.76 in)
Tyres		
Size (front and rear)	2.75 - 10 - 2PR	
Pressures:		
Front	21 psi (1.50 kg/cm²)	
Rear	25 psi (1.75 kg/cm²)	
Torque wrench settings	**lbf ft**	**kgf m**
Front wheel spindle retaining nut	29–36	4.0–50
Rear wheel retaining nut	58–72	8.0–10.0
Front and rear brake cam operating arm retaining nut	3.0–5.0	0.4–0.7

1 General description

The Honda Melody employs small diameter painted steel wheels of the type normally associated with scooters. The wheels are of pressed steel construction with an integral brake drum and wheel centre welded to the rim. The front wheel is of normal motorcycle fitting with a wheel spindle passing through the fork ends, whilst the rear wheel has a splined boss which engages a corresponding spline on the output shaft or stub axle of the reduction gearbox.

A simple single leading shoe (sls) drum brake is fitted to both wheels, the brakes being controlled by handlebar levers in a similar arrangement to that found on bicycles. Though similar in size and construction the wheels are not interchangeable.

Tyre sizes are 2.75-10-2PR front and rear and are of conventional tubed type.

2 Front wheel: examination and renovation

1 Place the machine on its centre stand and place a large wooden block, or similar, beneath the footrest panel mounting points so that the front wheel is raised clear of the ground.
2 Spin the wheel and check for rim alignment in both the axial and radial planes by placing a pointer close to the rim edge. If the total alignment variation is greater than the given service limit of 2.0 mm (0.08 in), then it is the manufacturer's recommendation that the wheel be renewed. This is, however, a counsel of perfection and in practice a larger amount of runout may not affect the handling qualities of the machine to an excessive degree.
3 When cleaning the machine, examine the wheel carefully. Any areas where the paint finish has been chipped away must be refinished at the earliest available opportunity if the likelihood of the wheel rusting is to be avoided. Any rust already present on the wheel should be removed by rubbing the affected area with a strip of emery cloth until the exposed metal is bright and clean and then coating the area with one of the rust removing agents sold by any of the well known motor accessory stockists before applying the final paint finish.
4 Inspect the complete wheel for localised damage in the form of cracks or dents. It is possible that a dent can be ignored as long as the paint finish has remained undamaged, but a crack will render the wheel unfit for further use unless it is found that a permanent repair is possible through welding. This method of repair requires a great degree of skill and therefore the advice of a wheel repair specialist should be sought.
5 Check closely that the area of wheel rim which supports the tyre bead is in no way damaged. A tyre that is improperly seated and which moves around the wheel rim can only constitute potential danger.
6 Finally, check the wheel for any signs of play in the wheel bearings. This may be indicated by being able to induce side-to-side movement of the wheel whilst the front forks are held securely in one position. Any roughness heard or felt in the bearings as the wheel is rotated will also mean that they are in need of immediate renewal.

3 Front wheel: removal and refitting

1 Place the machine on its centre stand and raise the front wheel clear of the ground by positioning a large wooden block, or similar, beneath the footrest panel mounting points.
2 Removal of the wheel is a simple and straightforward procedure which should begin with the detachment of the brake operating cable from the brake cam lever. To do this, wind the cable adjuster nut fully back so that there is the maximum amount of slack in the cable inner. Pull the cam lever rearwards and unhook the cable nipple from the lever end. The cable inner can now be eased through the slot in the brake backplate location for the cable adjuster and the cable allowed to hang clear of the wheel.
3 When disconnecting the brake cable, note the condition of the spring and rubber gaiter fitted over the cable end. The gaiter acts to protect the cable inner from contamination by road salts

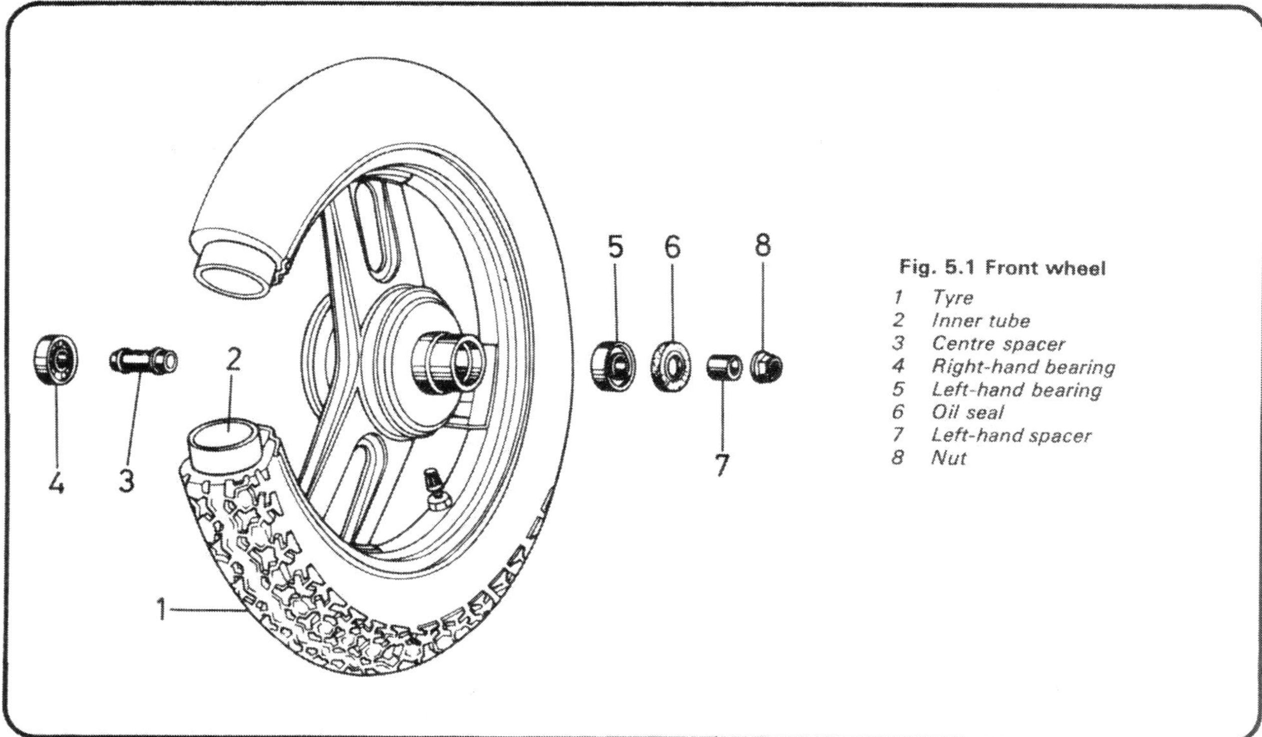

Fig. 5.1 Front wheel

1 *Tyre*
2 *Inner tube*
3 *Centre spacer*
4 *Right-hand bearing*
5 *Left-hand bearing*
6 *Oil seal*
7 *Left-hand spacer*
8 *Nut*

and dirt and therefore greatly lessens the likelihood of the cable becoming seized due to ingress of contamination between the cable inner and its outer. Renew the gaiter if it is split or perished. If the spring is seen to be broken or weakened by corrosion, then it also should be renewed.

4 Disconnect the speedometer drive cable from its location in the brake backplate by first removing the small cross head securing screw and then pulling the cable clear. Allow the cable to hang clear of the wheel.

5 Unscrew the flanged locknut from the end of the wheel spindle and carefully drift the spindle from position with the wheel supported. Use a soft-metal drift in conjunction with a hammer and take great care not to damage the threaded end of the spindle. The wheel can now be manoeuvred clear of the front forks.

6 Before fitting the wheel, clean the spindle of any hardened grease or corrosion and check it for straightness. Honda recommend that the spindle is supported between two V-blocks and its runout checked by means of a dial gauge positioned at the spindle centre. The runout measured should not exceed the service limit of 0.2 mm (0.008 in). It will suffice, however, to hold the spindle against a good straight-edge to check for straightness. A bent spindle must be replaced with a serviceable item. After having cleaned and checked the spindle, smear a light coating of a high melting point, lithium based grease along its length.

7 Fitting the front wheel is a direct reversal of the removal procedure, whilst noting the following points. When lifting the wheel into position between the forks, ensure that the projection of the right-hand pivot arm fits cleanly into the slot cast into the brake backplate. Note that if the brake backplate is allowed to rotate, due to its not being held in position by the projection of the pivot arm, the wheel will lock on the first application of the front brake, with disastrous consequences. The collar must be located in the oil seal.

8 Fit the wheel spindle with its retaining nut and tighten the nut to the specified torque loading of 29 – 36 lbf ft (4.0 – 5.0 kgf m). After having relocated the speedometer drive cable, secure it in position with the screw and then spin the front wheel whilst observing the instrument needle, to check that drive is being transmitted from the wheel to the instrument head.

9 Reconnect the brake cable and wind the adjuster nut forward until the amount of free play measured at the top of the handlebar lever is 10 – 15 mm (0.4 – 0.6 in). Check for correct operation of the brake by spinning the wheel and applying the brake lever. If, when the brake is off, the brake shoes are heard to be brushing against the surface of the wheel drum, back off the adjuster nut slightly until all indication of binding disappears. If the brake shoes have been renewed, then the brake may be readjusted after a period of bedding-in has been allowed.

3.7 The pivot arm projection must locate in its retaining slot (arrowed)

3.8a Tighten the spindle retaining nut to its specified torque loading

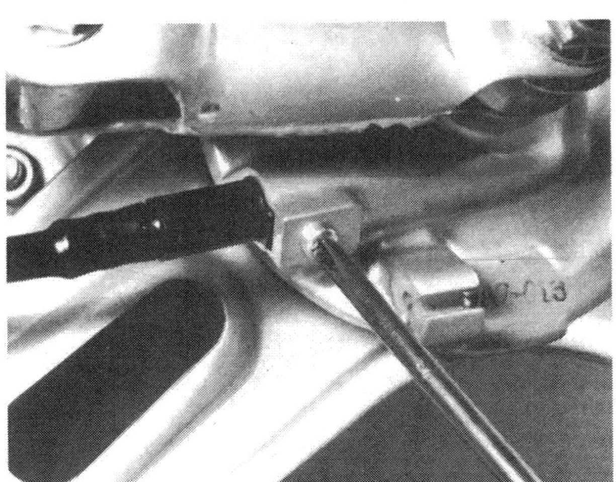
3.8b Secure the speedometer drive cable in position

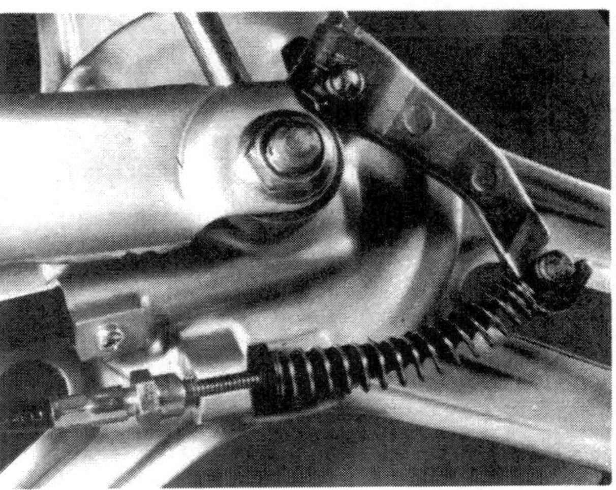
3.9 Reconnect the brake cable

4 Front wheel bearings: removal, examination and refitting

1 There are two bearings in the hub of the front wheel. If it is found possible to induce any side play in the wheel with it fitted to the machine, or to feel any roughness in the bearings as the wheel is rotated, then both wheel bearings need to be renewed.
2 Before bearing removal can be undertaken, the brake backplate assembly must be removed from the wheel hub and the wheel laid flat on a work surface, supported by wooden blocks so that enough clearance is left beneath the wheel to drive the bearing out. To lessen the risk of distortion to the wheel, ensure that these blocks are placed as close to the bearing as possible.
3 With the brake backplate side of the wheel hub facing uppermost, place the end of a small flat-ended drift against the upper face of the lower bearing and tap the bearing downwards out of the wheel hub. The spacer located between the two bearings may be moved sideways slightly in order to allow the drift to be positioned against the face of the bearing. Move the drift around the face of the bearing whilst drifting it out of position, so that the bearing leaves the hub squarely. The oil seal will be displaced along with the bearing. Note the spacer collar located in the centre of this seal.
4 With the one bearing removed, the wheel may be lifted and the spacer withdrawn from the hub. Invert the wheel and remove the second bearing, using a similar procedure to that used for the first.
5 Wash both bearings, the bearing spacer and their hub location thoroughly in clean petrol to remove all traces of the old grease. Remember to take the necessary fire precautions whilst doing this. Check the bearing tracks and balls for wear, pitting or damage to the hardened surfaces. A small amount of side movement in the bearing is normal but no radial movement should be detectable. If wear or damage is found the bearing in question should be renewed.
6 If the original bearings are to be refitted, then they should be repacked with the recommended grease before being fitted into the hub. New bearings must also be packed with the recommended grease. Ensure that the bearing recesses in the hub are clean and both bearing and recess mating surfaces lightly greased. The two bearings and central spacer may now be fitted. With the hub well supported by the wooden blocks, drift the bearing into the brake backplate side of the wheel hub using a length of metal tube, or a socket, of suitable diameter and a soft-faced mallet. Invert the wheel, insert the spacer and fit the second bearing using the same procedure as that given for the first. Note that both bearings must be fitted with their sealed sides facing outwards.
7 It is considered good practice to renew the oil seal every time the bearings are removed. In any event, the seal must be renewed if it is seen to be damaged or if it has begun to deteriorate. With the seal pressed partially into its recess in the wheel hub, tap it home by using a socket or length of metal tube of suitable diameter in conjunction with a soft-faced hammer. Lightly grease the spacer collar and insert it into position through the seal.

5 Speedometer drive gear: examination and renovation

1 The speedometer drive assembly is contained within the front wheel brake backplate and should be examined and repacked with grease whenever work is carried out on the wheel bearings or brake assembly.
2 To remove the main drive gear from its housing, place the brake backplate assembly, inner side uppermost, on a work surface. Remove the brake shoes and spring, as described in Section 10 of this Chapter, and inspect the large dust seal for signs of damage and deterioration. To renew this seal, carefully lever it out of position using the flat of a screwdriver; great care

4.6a Do not omit to refit the bearing spacer

4.6b Each bearing must be fitted with its sealed side facing outwards ...

4.6c ... before it is drifted into position

4.7a Fit a serviceable oil seal ...

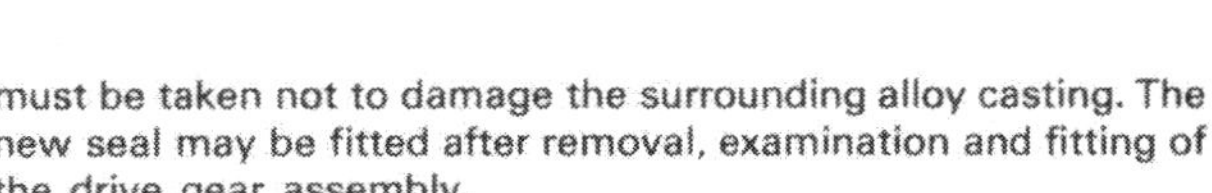

4.7b ... and insert the greased spacer collar through it

must be taken not to damage the surrounding alloy casting. The new seal may be fitted after removal, examination and fitting of the drive gear assembly.

3 The main drive gear can now be withdrawn from its housing. In practice, some difficulty may be experienced in withdrawing the gear due to it being engaged with the worm drive gear and to the retentive qualities of the grease around the base of the gear.

4 Remove all old grease from the main drive gear and from its housing by wiping the components with a clean rag. It is unlikely that the main drive gear will wear to any excessive degree until the machine has covered a considerable number of miles. Nevertheless, inspect the gear for broken teeth, broken drive tabs or any sign of excessive wear. If the gear is seen to be defective, then it must be renewed. Carry out a similar inspection on the worm drive gear. Unfortunately, Honda do not list this component as a separate item to the brake backplate so it must be assumed that any defective worm drive gear will necessitate replacement of the complete backplate. Return the component to an official Honda service agent for further advice on this matter.

5 To reassemble the speedometer drive gear, pack the housing with a high melting point, lithium based grease before inserting the main drive gear into position. Where required, fit the new dust seal into its recess in the brake backplate by pressing it into position evenly and squarely. Wipe a small amount of grease over the lip of this seal. The brake shoes and spring may now be refitted and the brake backplate assembly inserted into the wheel hub. Take care to ensure that the drive tabs of the main drive gear are aligned correctly with the corresponding slots in the wheel hub boss.

6 Rear wheel: examination and renovation

1 Place the machine on its centre stand and position a support beneath the engine crankcase to keep the rear wheel from coming into contact with the ground. With an assistant steadying the machine, check for wear in the wheel centre spline by gripping each side of the wheel and attempting to move it from side-to-side. Any movement between the wheel and its stub-axle will necessitate removal of the wheel for further investigation.

2 The remaining examination and renovation procedure is as for the front wheel (see Section 2, paragraphs 2 to 5). Rotate the wheel to check for roughness in the stub axle bearings.

5.2 Remove the speedometer drive gear seal ...

5.3 ... and withdraw the main drive gear from its housing

7 Rear wheel: removal and refitting

1 Commence removal of the rear wheel by placing the machine securely on its centre stand and then removing the exhaust system by following the instructions given in Section 16 of Chapter 2.

2 It is now necessary to prevent the rear wheel from rotating directly any turning force is imposed on the wheel securing nut in an attempt to loosen it. This is most easily accomplished by taking the machine off its stand and having an assistant sit astride it whilst at the same time applying both brakes to prevent the machine from moving forward. If an assistant is not readily available, it may be possible to lock the wheel sufficiently by turning the brake cable adjuster fully in so that the brake is applied. The wheel securing nut is tightened to a high torque loading and will therefore be difficult to loosen.

3 Remove the wheel securing nut and the thick plate washer located beneath it. Position a support beneath the engine crankcase so that the back of the machine is prevented from dropping, grip each side of the wheel and pull it clear of its stub-axle and away from the machine. Note the presence of the thrust collar contained within the recess of the transmission housing.

4 With the wheel removed, take the opportunity to clean and inspect the splines of the stub-axle and of the wheel itself. Inspect the load bearing surfaces of each spline for signs of deterioration in its surface finish. This, and any wear lip found along the length of each spline, will necessitate renewal of both the shaft and the wheel. Carry out a final check for wear by temporarily refitting the wheel over the shaft and feeling for any movement between the two components. With both shaft and wheel considered to be serviceable, apply a very light smear of a high melting point, lithium based grease to the shaft spline before finally sliding the wheel over the shaft. Bear in mind that just enough grease is needed to prevent dryness and the resulting ingress of moisture between the spline surfaces. There is a distinct possibility that over lubrication will lead to contamination of the brake shoes.

5 It is considered good practice to renew the wheel securing nut each time it is removed. This is because every locknut of this type loses some of its gripping qualities each time it is moved along the threaded section of shaft to which it is attached and at some stage it is going to unwind whilst the machine is in motion, with the resulting disastrous consequences. Bear in mind that the price of a single locknut is nothing in comparison to that of a new transmission casing. If it is found impossible to obtain a nut, then coat the threads of the shaft with a thread locking compound before fitting the nut with its washer and tightening it to the recommended torque loading of 58 – 72 lbf ft (8.0 – 10.0 kgf m).

6 Refit the exhaust system by following the instructions given in Section 16 of Chapter 2. Before riding the machine, carry out a quick check to ensure that the rear brake is correctly adjusted. There should be 10 – 15 mm (0.4 – 0.6 in) of free play measured at the tip of the handlebar lever. The amount of free play can be adjusted by rotation of the adjuster nut at the wheel end of the cable. If the brake shoes are heard to be brushing against the surface of the wheel drum as the wheel is rotated, then back off the adjuster nut slightly until all indication of binding disappears. If the brake shoes have been renewed, then the brake may be readjusted after a period of bedding-in has been allowed. Check that the brake works efficiently before taking the machine on the road.

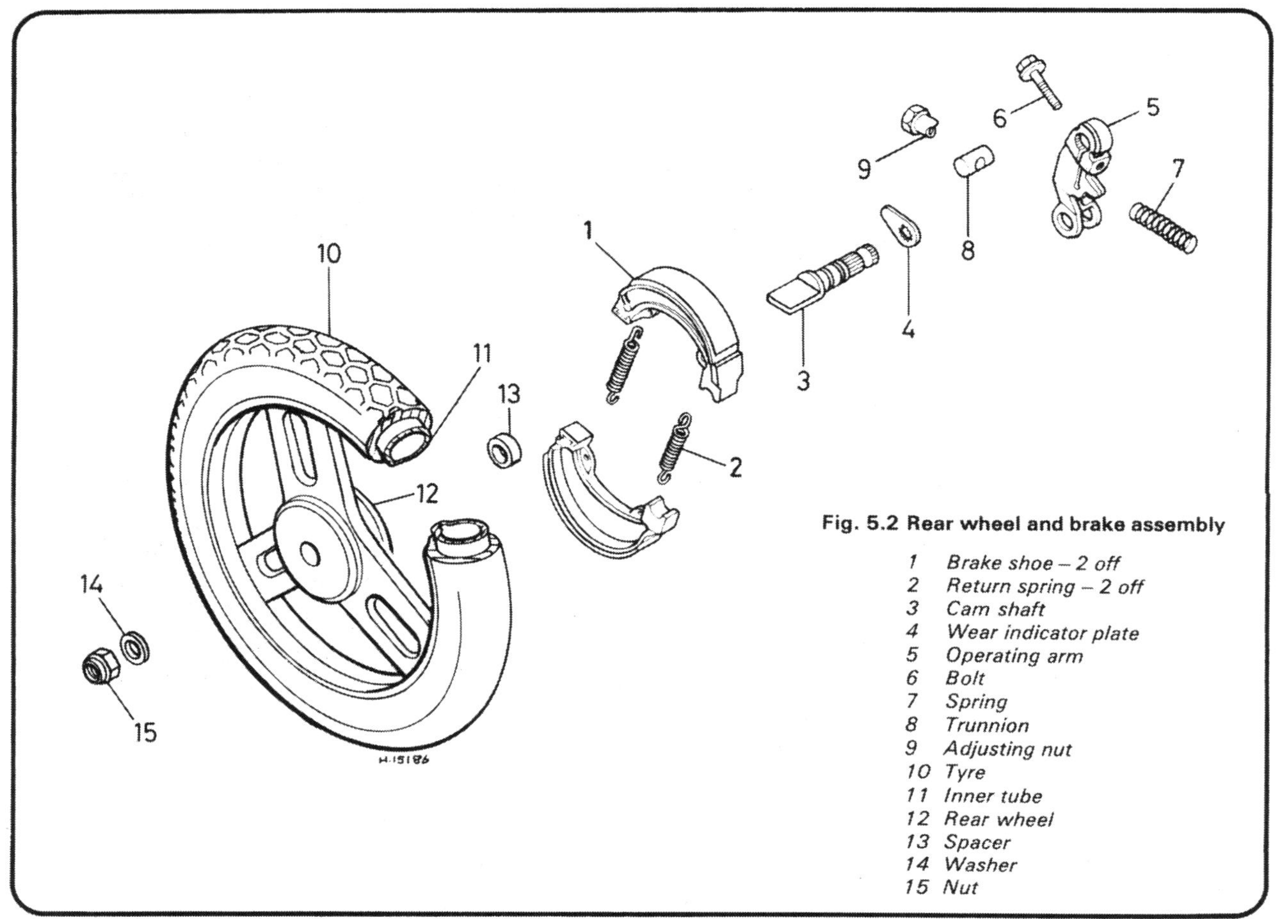

Fig. 5.2 Rear wheel and brake assembly

1 Brake shoe – 2 off
2 Return spring – 2 off
3 Cam shaft
4 Wear indicator plate
5 Operating arm
6 Bolt
7 Spring
8 Trunnion
9 Adjusting nut
10 Tyre
11 Inner tube
12 Rear wheel
13 Spacer
14 Washer
15 Nut

Tyre changing sequence - tubed tyres

A — Deflate tyre. After pushing tyre beads away from rim flanges push tyre bead into well of rim at point opposite valve. Insert tyre lever adjacent to valve and work bead over edge of rim.

B — Use two levers to work bead over edge of rim. Note use of rim protectors

C — Remove inner tube from tyre

D — When first bead is clear, remove tyre as shown

E — When fitting, partially inflate inner tube and insert in tyre

F — Work first bead over rim and feed valve through hole in rim. Partially screw on retaining nut to hold valve in place.

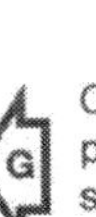

G — Check that inner tube is positioned correctly and work second bead over rim using tyre levers. Start at a point opposite valve.

H — Work final area of bead over rim whilst pushing valve inwards to ensure that inner tube is not trapped

8 Rear wheel bearings: general information

1 Unlike the front wheel, the rear wheel is not fitted with its own bearings. The hub is splined and fits directly on the rear stub axle, this being carried on bearings in the final drive casing. If play is detected, it will be necessary to examine these bearings for wear or damage, and reference should be made to Chapter 1 for details.

9 Brakes: adjustment and checking for wear

1 It is essential that the brakes on any motorcycle are always kept in correct adjustment. The procedure for carrying out adjustment of the brakes on the Honda Melody is both simple and straightforward and identical for each brake. Commence by checking the amount of free play measured at the handlebar lever end. If the amount of play measured is outside the set limit of 10 – 15 mm (0.4 – 0.6 in), then it must be altered by turning the adjuster nut at the brake drum end of the cable. Check for correct operation of the brake by spinning the wheel and applying the brake lever. If, when the brake is off, the brake shoes are heard to be brushing against the surface of the wheel drum, then back off the adjuster nut slightly until all indication of binding disappears. If the brake shoes have been renewed, then the brake should be readjusted after a period of bedding-in has been allowed.
2 Honda state that for the front brake only, it is permissible to compensate for wear on the brake shoe linings by moving the brake cam operating arm in relation to the cam itself. If it is found that it is no longer possible to obtain the correct amount of free play at the handlebar lever and by turning the cable adjuster, then the brake cam operating arm should be detached from the cam shaft and moved one spline in an anti-clockwise direction before being refitted. Note that this operation may only be performed once.
3 With the brakes correctly adjusted, apply each handlebar lever fully and check the position of the indicator plate mounted on the brake cam shaft with the arrow cast into the brake backplate. If the arrow on the indicator plate is seen to align with or go past the arrow on the brake backplate, then the brake shoe linings have worn beyond their service limits and should be renewed immediately.

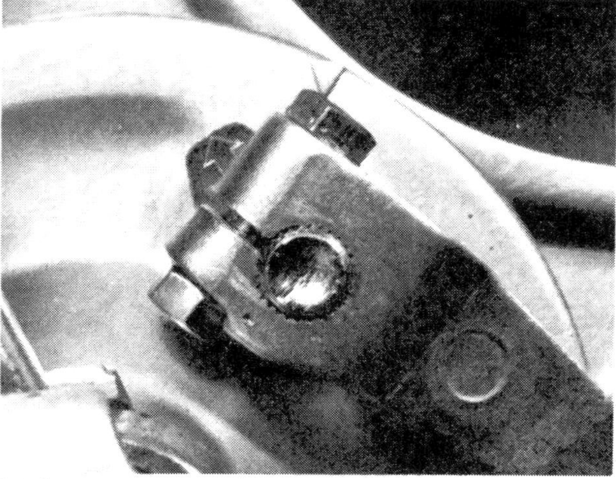
9.3 Check the position of the wear indicator plate

10 Brakes: examination and renovation

1 Both wheels of the Honda Melody incorporate a single leading shoe (sls) drum brake, each of which is operated by a handlebar-mounted lever.
2 To gain access to the components of each brake assembly, it is first necessary to remove the appropriate wheel as described in Sections 3 or 7 of this Chapter. In the case of the front wheel, the brake backplate assembly is removed complete with the wheel from the machine; it can then be easily pulled clear of the wheel hub. The components of the rear brake assembly will remain attached to the transmission casing of the engine/transmission unit.
3 The brake drums should be checked for wear or damage after any accumulation of dust has been removed. It should be noted that this dust contains asbestos which can be harmful if inhaled. For this reason, **never** use compressed air to remove the dust. A petrol moistened rag will remove the dust and clean the drum surface quickly and safely. Look for signs of scoring or any other damage on the drum surface. If badly damaged it may be possible to have the drum skimmed in a lathe by a suitably-equipped specialist. Failing this, renewal of the entire wheel is inevitable. In practice, such damage is unlikely to be encountered, and very light scoring can be accepted. Any rusting of the drum surface may be removed by careful use of fine abrasive paper.
4 The shoes can be checked for wear by measuring the thickness of the lining material at the point where the greatest amount of wear has taken place. The service limit for the thickness of the lining material is 2.0 mm (0.08 in).
5 The front brake shoes can be removed once the return spring has been released. This takes the form of a large C-shaped spring ring, the ends of which engage in holes in the shoe ends. To release the shoes, grasp the ends furthest from the brake cam, pulling them apart and away from the backplate. The shoes can be allowed to fold inwards to help in freeing the brake cam ends. It is recommended that eye protection is worn during this operation because the spring can jump free with unexpected force in an unpredictable direction.
6 The rear brake shoes have conventional extension coil springs fitted between the shoes. They can be freed by folding the shoes inwards until the ends can be disengaged from the cam and pivot. Once the shoes have been removed note the position of each spring in relation to the brake shoes before separating them.
7 If the existing shoes are to be refitted, they should be wiped with a clean petrol-moistened rag to remove the accumulated brake dust. The rear shoes in particular should be examined for signs of oil contamination. If this is slight it may be possible to remove the oil with petrol. An aerosol solvent known as Ultraclean, used mostly for electrical cleaning and degreasing, is ideal for this job and can be obtained from motor factors or through electrical shops. Severe contamination will necessitate renewal of the shoes. It follows that where oil contamination has been noted it will be necessary to rectify the leak before the brake is refitted. This is most likely to be a damaged oil seal on the stub axle and reference should be made to Chapter 1 for details.
8 Existing shoes that are to be refitted should also have their lining surface roughened sufficiently to break the glaze which will have formed in use. Glasspaper or emery cloth is ideal for this purpose but take great care not to inhale any of the asbestos dust that may come from the lining surface.
9 Before fitting the brake shoes, check that the brake operating cam is working smoothly and not binding in its pivot. The cam may be removed by unscrewing the operating arm retaining bolt and pulling the arm off the shaft. The cam shaft can then be extracted from its location in the brake backplate. Thoroughly clean the shaft and the bore in the brake backplate through which it passes. Clean and grease the shaft prior to reassembly and also lightly grease the faces of the operating

cam. A light smear of grease should also be placed on the face of the spigot about which the shoes pivot. The brake shoe wear indicator plate is aligned with the cam shaft by matching the wide groove in the shaft end with the wide tooth of the plate. When fitting the cam operating arm, align the punch mark on the arm with that on the shaft. Note that the cam shaft of the front brake incorporates a dust seal. If this seal is seen to be defective then it must be renewed.

10 Finally, inspect the brake shoe springs for signs of damage or deterioration. After a considerable amount of use, the springs will take a 'set' and become inefficient. If it is suspected that this has happened, take the spring in question to an official Honda service agent and compare it with a new item. Note that the two springs of the rear brake must be of identical lengths and must always be renewed as a pair.

11 Check the spring to brake shoe contact points. If either component is excessively worn, then it must be renewed. If the ingress of moisture into the brake drums has caused the springs to become corroded and heavily pitted, then they should be renewed before subsequent failure occurs. Do not neglect to examine the rear brake cam operating arm return spring by using a procedure similar to that described above.

12 Reassemble and refit the brake assemblies by reversing the dismantling and removal procedures. With the rear brake assembly, it was found in practice that holding the shoes in the V shape ready to fit them onto the backplate and at the same time retaining the return springs, required a little practice and a lot of patience. Once the ends of the shoes were located with the cam and pivot points however, it was easy to snap them into position by pressing downward. On no account use excessive force, otherwise there is a risk that the shoes may be permanently distorted, or the springs over-stretched. Do not forget to reset the brake adjustment or to test the brakes for correct operation before taking the machine on the road.

10.2 The front wheel rake assembly can be pulled clear of the wheel hub

10.10 Do not neglect the rear brake arm return spring

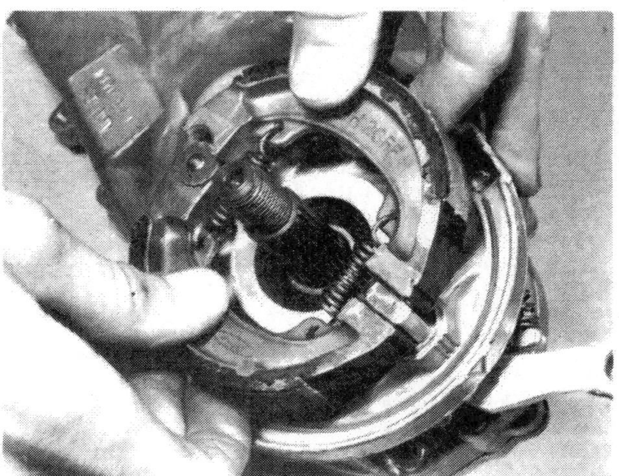
10.12 Do not use excessive force when relocating the rear brake shoes

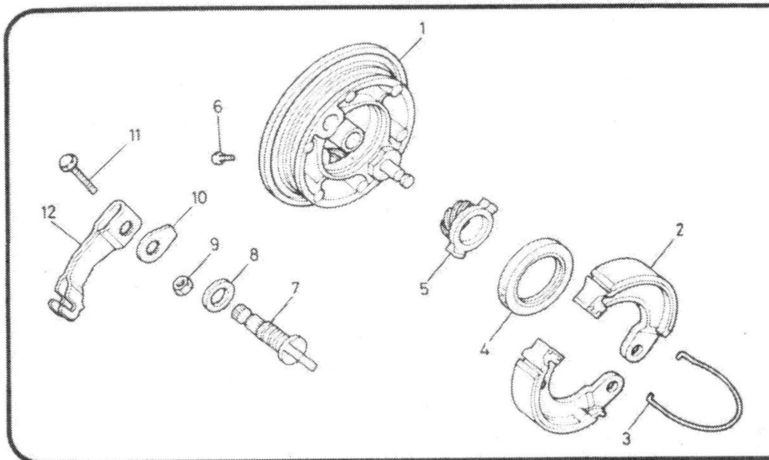

Fig. 5.3 Front brake assembly

1 *Backplate*
2 *Shoes*
3 *Return spring*
4 *Oil seal*
5 *Speedometer drive gear*
6 *Screw*
7 *Cam shaft*
8 *Dust seal*
9 *Nut*
10 *Wear indicator plate*
11 *Bolt*
12 *Operating arm*

11 Tyres: removal and refitting

1 At some time or other the need will arise to remove and replace the tyres, either as a result of a puncture or because a replacement is required to offset wear. To the inexperienced, tyre changing represents a formidable task, yet if a few simple rules are observed and the technique learned, the whole operation is surprisingly simple.

2 To remove the tyre from either wheel, first detach the wheel from the machine by following the procedure in Sections 3 or 7 of this Chapter, depending on whether the front or the rear wheel is involved. Deflate the tyre by removing the valve insert and when it is fully deflated, push the bead of the tyre away from the wheel rim on both sides so that the bead enters the centre well of the rim. Remove the locking cap and push the tyre valve into the tyre itself.

3 Insert a tyre lever close to the valve and lever the edge of the tyre over the outside of the wheel rim. Very little force should be necessary; if resistance is encountered it is probably due to the fact that the tyre beads have not entered the well of the wheel rim all the way round the tyre.

4 Once the tyre has been edged over the wheel rim it is easy to work around the wheel rim so that the tyre is completely free on one side. At this stage, the inner tube can be removed.

5 Working from the other side of the wheel, ease the other edge of the tyre over the outside of the wheel rim that is furthest away. Continue to work around the rim until the tyre is free completely from the rim.

6 If a puncture has necessitated the removal of the tyre, reinflate the inner tube and immerse it in a bowl of water to trace the source of the leak. Mark its position and deflate the tube. Dry the tube and clean the area around the puncture with a petrol-soaked rag. When the surface has dried, apply the rubber solution and allow this to dry before removing the backing from the patch and applying the patch to the surface.

7 It is best to use a patch of the self-vulcanising type, which will form a very permanent repair. Note that it may be necessary to remove a protective covering from the top surface of the patch after it has sealed in position. Inner tubes made from synthetic rubber may require a special type of patch and adhesive, if a satisfactory bond is to be achieved.

8 Before replacing the tyre, check the inside to make sure the agent that caused the puncture is not trapped. Check also the outside of the tyre, particularly the tread area, to make sure nothing is trapped that may cause a further puncture.

9 If the inner tube has been patched on a number of past occasions, or if there is a tear or large hole, it is preferable to discard it and fit a replacement. Sudden deflation may cause an accident.

10 To replace the tyre, inflate the inner tube sufficiently for it to assume a circular shape but only just. Then push it into the tyre so that it is enclosed completely. Lay the tyre on the wheel at an angle and insert the valve through the rim tape and the hole in wheel rim. Attach the locking cap on the first few threads, sufficient to hold the valve captive in its correct location.

11 Starting at the point furthest from the valve, push the tyre bead over the edge of the wheel rim until it is located in the central well. Continue to work around the tyre in this fashion until the whole of one side of the tyre is on the rim. It may be necessary to use a tyre lever during the final stages.

12 Make sure there is no pull on the tyre valve and again commencing with the area furthest from the valve, ease the other bead of the tyre over the edge of the rim. Finish with the area close to the valve, pushing the valve up into the tyre until the locking cap touches the rim. This will ensure the inner tube is not trapped when the last section of the bead is edged over the rim with a tyre lever.

13 Check that the inner tube is not trapped at any point. Reinflate the inner tube, and check that the tyre is seating correctly around the wheel rim. There should be a thin rib moulded around the wall of the tyre on both sides, which should be equidistant from the wheel rim at all points. If the tyre is unevenly located on the rim, try bouncing the wheel when the tyre is at the recommended pressure. It is probable that one of the beads has not pulled clear of the centre well.

14 Always run the tyres at the recommended pressures and never under or over-inflate. The correct pressures for solo use are given in the Specifications Section of this Chapter. It should be remembered that the small size of the tyres means that the loss of a small quantity of air will result in a significant drop in pressure, so regular checks on tyre pressure should not be overlooked.

15 Tyre replacement is aided by dusting the side walls, particularly in the vicinity of the beads, with a liberal coating of French chalk. Washing-up liquid can also be used to good effect, but this has the disadvantage of causing the inner surfaces of the wheel to rust.

16 Never fit a tyre that has a damaged tread or side wall. Apart from the legal aspects, there is a very great risk of a blow-out which can have serious consequences on any two wheel vehicle.

17 Tyre valves rarely give trouble, but it is always advisable to check whether the valve itself is leaking before removing the tyre. Do not forget to fit the dust cap, which forms an effective second seal. The tyre valve dust cap is often left off when a tyre has been replaced, despite the fact that it serves an important two-fold function. Firstly, it prevents dirt or other foreign matter from entering the valve and causing the valve to stick open when the tyre pump is next applied. Secondly, it forms an effective second seal so that in the event of the tyre valve leaking, air will not be lost.

12 Fault diagnosis: wheels, brakes and tyres

Symptom	Cause	Remedy
Handlebars oscillate at low speeds	Buckle or flat in wheel rim, most probably front wheel Tyre not straight on rim	Check rim alignment by spinning wheel. Renew wheel if damaged. Check tyre alignment.
Machine lacks power and accelerates poorly	Brakes binding	Warm brake drums provide best evidence. Re-adjust brakes.
Brakes grab when applied gently	Ends of brake shoes not chamfered Elliptical brake drum	Chamfer with file. Lightly skim in lathe (specialist attention needed).
Brake pull-off sluggish	Brake cam binding in housing Weak brake shoe springs	Free and grease. Renew if springs not displaced.
Brakes ineffective	Contaminated or glazed linings	Remove and renew or remove glaze as necessary.
Brakes feel spongy	Cable badly routed Stretched brake operating cables	Re-route cable(s) avoiding sharp bends. Renew cables.
Tyre wears more rapidly in middle of tread	Over inflation	Check pressures and run at recommended settings.
Tyres wear rapidly at outer edges of tread.	Under inflation	Ditto.

Chapter 6 Electrical system

Refer to Chapter 7 for information on the NB, ND and NP50 models

Contents

Specifications

Electrical system

Type	Flywheel generator, direct lighting
Voltage	12V
Earth	Negative
Generator output	93W @ 5000 rpm

Charging system

Output	Lights off	Lights on
At 1800 rpm or less	12.6V	–
At 2000 rpm or less	–	12.6V
At 2500 rpm	14.8V (0.3A) or more	14.8V (0.3A) or more
At 6000 rpm	15.5V (1.0A) or less	15.5V (2.0A) or less

Battery

Make	Yuasa
Type	YB4L-B
Voltage	12V
Capacity	4 Ah
Electrolyte specific gravity	1.270 – 1.290 at 20°C (68°F)
Maximum charging rate	0.4A

Fuse

Fuse	7A

Bulbs

Headlamp	12V 25/25W
Tail/stop lamp	12V 5/21W
Pilot lamp	12V 3.4W
Indicator lamps	12V 21W
Oil level warning light	12V 3.4W
Indicator warning light	12V 3.4W
Speedometer light	12V 3.4W

1 General description

The Honda Melody features a 12 volt negative earth electrical system. Power to the system comes from a charging coil which forms part of the crankshaft mounted flywheel generator assembly. The alternating current (ac) output from the generator is converted to direct current (dc) by a silicon rectifier and then passed on via a 7 amp fuse to provide a charging current for the battery.

Power required to illuminate the headlamp, tail lamp, pilot lamp and speedometer light comes direct from the flywheel generator in the form of alternating current. A resistor is included in the circuit for the purpose of absorbing power from the generator whilst these lights are switched off.

The battery has a 4 ampere hour rating and provides direct current to illuminate the direction indicator lamps, together with the stop lamp, direction indicator warning light, oil level warning light and fuel level indicator. Other components operated by the battery include the fuel and oil level indicator switches, the horn and the direction indicator relay.

Honda Melody deluxe models (the NS50 MS) are fitted with a starter motor as an alternative means of starting the engine to the kickstart which is fitted as standard to all Melody models. This motor is operated by a push button which is mounted on the right-hand handlebar but will not function unless the lever which operates the rear brake is locked in the 'On' position. Power to turn the motor is provided by the battery in the form of direct current; this current being passed through a starter relay which is energised by the push button being pressed.

2 Testing the electrical system: general information

1 The electrical system incorporated in the Honda Melody lends itself to fairly comprehensive testing of its component parts. A certain amount of preliminary dismantling is necessary to gain access to the components to be tested and full information on removing and fitting the various body panels mentioned in the following Sections of this Chapter is given in Chapter 4 of this Manual.

2 Simple continuity checks may be made using a dry battery and bulb arrangement, but for most of the tests in this Chapter a pocket multimeter can be considered essential. Many owners will already possess one of these devices, but if necessary they can be obtained from electrical specialists or mail order companies.

3 Care must be taken when performing any electrical test, because some of the electronic assemblies can be destroyed if they are connected incorrectly or inadvertently shorted to earth. Instructions regarding meter probe connections are given for each test, and these should be read carefully to preclude any accidental damage during the test. Note that separate amp, volt and ohm meters may be used in place of the multimeter if necessary, noting that the appropriate test ranges will be required.

4 Where test equipment is not available, or the owner feels unsure of the procedure described, it is recommended that professional assistance is sought. Do not forget that a simple error can destroy a component such as the rectifier, resulting in expensive replacements being necessary.

3 Wiring: layout and examination

1 The wiring harness is colour coded and will correspond with the accompanying diagram. Where socket connectors are used they are designed so that reconnection can be made only in the one correct position.

2 Visual inspection will show whether any breaks or frayed outer coverings are giving rise to short circuits. Another source of trouble may be the snap connectors and sockets, where the connector has not been pushed home fully in the outer housing.

3 Intermittent short circuits can often be traced to a chafed wire that passes through or is close to a metal component, such as a frame member. Avoid tight bends in the wire or situations where the wire can become trapped between casings.

4 Charging system: checking the output

1 Power is supplied to the charging system of the machine by a coil which forms part of the crankshaft mounted flywheel generator assembly. Components included in the system are the battery, a 7 amp fuse and a silicon rectifier.

2 Before attempting to check the output of the charging system as a whole, ensure that the battery is in good condition and fully charged. Start the engine and run it until it has reached its normal operating temperature. Turn the engine off, lift the seat of the machine and connect an ammeter and a voltmeter to the points shown in the figure accompanying this text.

3 Restart the engine and check the ammeter and voltmeter readings, comparing them with the figures shown in the table below. Unless a tachometer is available, the engine speed will have to be estimated; this should, however, suffice to give an impression of the charging system condition. Note that it must be remembered that the transmission will become engaged as the engine speed increases, so make quite sure that the rear wheel is raised clear of the ground and that there is no risk of the machine rolling off its stand.

Charging system output	Lights off	Lights on
At 1800 rpm or less	12.6V	–
At 2000 rpm or less	–	12.6V
At 2500 rpm	14.8V (0.3A) or more	14.8V (0.3A) or more
At 6000 rpm	15.5V (1.0A) or less	15.5V (2.0A) or less

4 If the readings obtained are substantially different from those shown above, then check the condition of the component parts in the system by carrying out the checks listed in the following Sections of this Chapter. Before doing this, take note of the information given in Section 3 of this Chapter in relation to the examination of the wiring between the component parts.

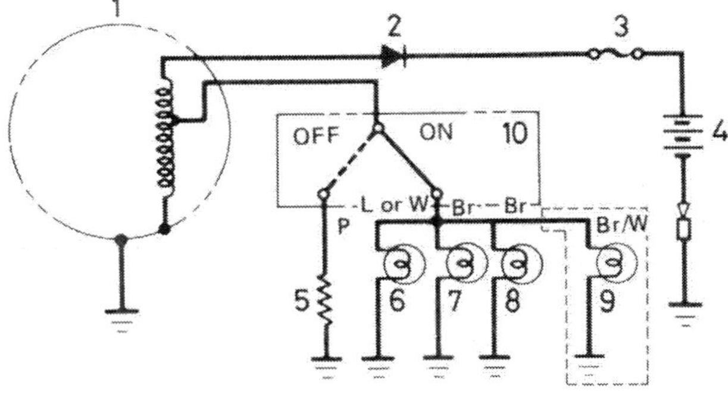

Fig. 6.1 Charging system circuit diagram

1 Flywheel generator
2 Silicon rectifier
3 7 amp fuse
4 Battery
5 Resistor
6 Headlamp
7 Speedometer lamp
8 Tail lamp
9 Pilot lamp
10 Lighting switch

5 Charging/lighting coil: testing and renewal

1 The charging/lighting coil is mounted on the stator plate of the flywheel generator assembly and is instrumental in providing a charging current to the battery and in providing power to illuminate the headlamp, tail lamp, pilot lamp and speedometer light. If it is suspected that failure of this coil has occurred, then it may be quickly and easily checked by carrying out the following instructions.
2 Remove the centre body panel from the machine to reveal the electrical wires which are routed down the left-hand side of the frame structure, just forward of the cylinder head. Unplug the push connectors of the Yellow and White wires. Set a multimeter to its resistance function (x1 ohm) and measure the resistance between the Yellow wire and a good earth point on the crankcase. If the reading obtained is between the limits 0.3 – 1.5 ohm, then the lighting section of coil is serviceable. Check that the resistance between the White wire and the earth point is between 0.5 – 1.5 ohm. If this reading is correct, then the charging system section of coil is serviceable.
3 If the coil is found to be unserviceable in either one of its sections, then the complete flywheel generator assembly will have to be renewed as Honda do not list its component parts as separate items. Always have a Honda service agent confirm that the coil is defective before going to the expense of purchasing a replacement item. Remember that various alternatives to purchasing a new item are available; these include obtaining advice from a competent auto-electrician as to whether the coil can be repaired and contacting the various 'bike breakers' who advertise in the weekly and monthly motorcycle journals. Removal and fitting of the flywheel generator assembly are dealt with in the relevant Sections of Chapter 1.

5.1 The charging coil is mounted on the stator plate of the flywheel generator

6 Silicon rectifier: location and testing

1 The silicon rectifier is included in the charging system of the machine and serves to convert the alternating current (ac) output from the flywheel generator to direct current (dc) for the purpose of charging the battery. This it does by blocking half of the output wave from the generator. Its operation is analogous to that of a one way valve in that it will allow the current to flow in one direction only.
2 Access to the rectifier can be gained by removal of the centre body panel from the machine; this panel being held in position by a dome head nut with washer and a single screw. The rectifier is located on the right-hand side of the machine, just below the fuel tank and directly in front of the starter solenoid switch and flasher unit.
3 To test the rectifier, disconnect the push connector from its base and set a multimeter to its resistance function. Connect one probe of the meter to each of the two rectifier terminals. Note the reading obtained, then reverse the probe leads and check the reading once more. If the rectifier is functioning normally it should allow current to pass in one direction but not in the other, thus a reading of no resistance should be shown, with infinite resistance when the probes are reversed. If the readings obtained do not correspond, the unit is faulty and must be renewed.

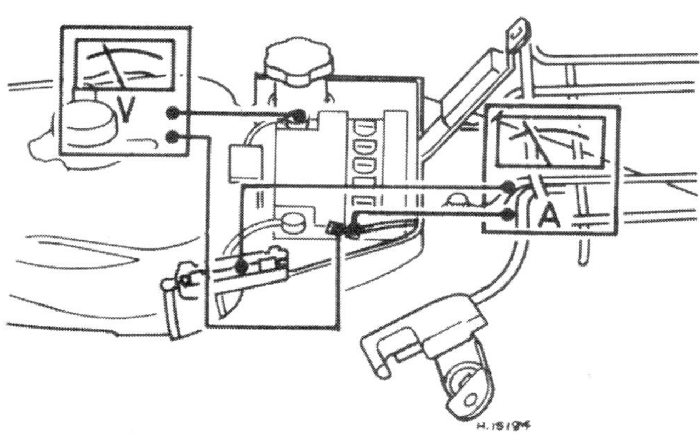

Fig. 6.2 Charging system test

7 Fuse: location, function and renewal

1 The fuse is of a 7 amp rating and is contained within a plastic holder which is normally located in a small casing which forms part of the left-hand side of the battery tray.
2 The fuse is fitted to protect the electrical system in the event of a short circuit or sudden surge. It is, in effect, an intentional 'weak link' which will blow in preference to the circuit burning out.
3 Before replacing a fuse that has blown, check that no obvious short circuit has occurred, otherwise the replacement fuse will blow immediately it is inserted. It is always wise to check the electrical circuit thoroughly, to trace the fault and eliminate it.
4 When a fuse blows while the machine is running and no spare is available, a 'get you home' remedy is to remove the blown fuse and wrap it in silver paper before replacing it in the fuseholder. The silver paper will restore the electrical continuity by bridging the broken fuse wire. This expedient should **never** be used if there is evidence of a short circuit or other major electrical fault, otherwise more serious damage will be caused. Replace the 'doctored' fuse at the earliest possible opportunity, to restore full circuit protection. It follows that spare fuses that are used should be replaced as soon as possible to prevent the above situation from arising.

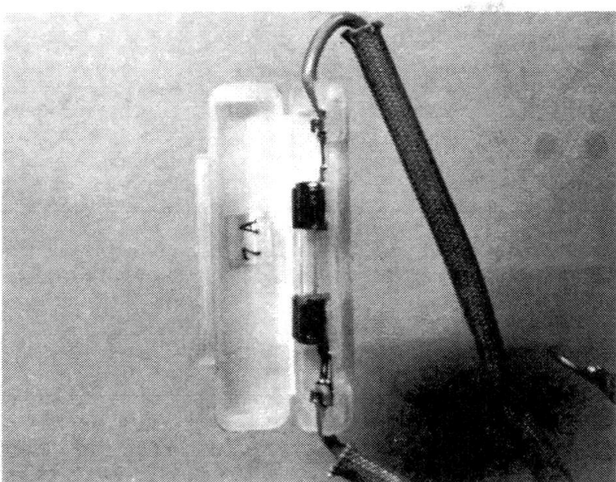

7.1 The fuse is contained within a plastic holder

8 Battery: examination and maintenance

1 A Yuasa YB4L-B battery is fitted as standard to the Honda Melody and is of the lead-acid type. The translucent plastic case of the battery permits the upper and lower levels of the electrolyte to be observed when the battery is lifted out of its tray. Maintenance is normally limited to keeping the electrolyte level between the prescribed upper and lower limits and by making sure the vent pipe is not blocked and remains correctly routed.
2 Unless acid is spilt, as may occur if the machine falls over, the electrolyte should always be topped up with distilled water, to restore the correct level. If acid is split on any of the machine, it should be neutralised with an alkali such as washing soda and washed away with plenty of water, otherwise serious corrosion will occur. Top up with sulphuric acid of the correct specific gravity (1.270 – 1.290) only when spillage has occurred. Check that the vent pipe is well clear of the frame structure or any of the cycle components, for obvious reasons.
3 If battery problems are experienced, the following checks will determine whether renewal is required. A battery can normally be expected to last for about 3 years, but this life can be shortened dramatically by neglect. In normal use, the capacity for storage will gradually diminish, and a point will be reached where the battery is adequate for all but the strenuous task of starting the engine.
4 Remove the flat battery and examine the cell and plate condition near the bottom of the casing. An accumulation of white sludge around the bottom of the cells indicates sulphation, a condition which indicates the imminent demise of the battery. Little can be done to reverse this process, but it may help to have the electrolyte drained, the battery flushed and then refilled with new electrolyte. Most electrical wholesalers have facilities for this work.
5 Warping of the plates or separators is also indicative of an expiring battery, and will often be evident in only one or two of the cells. It can often be caused by old age, but a new battery which is overcharged will show the same failure. There is no cure for the problem and the need to avoid overcharging cannot be overstressed.
6 Try charging the suspect battery as described in the following Section. If the battery fails to accept a full charge, and in particular, if one or more cells show a low hydrometer reading, the battery is in need of renewal.
7 A hydrometer will be required to check the specific gravity of the electrolyte, and thus the state of charge. Any small hydrometer will do, but avoid the very large commercial types because there will be insufficient electrolyte to provide a reading. When fully charged, each cell should read 1.270 – 1.290, with little discrepancy between cells.
8 Note that it is seldom practicable to repair a cracked battery case because the acid present in the joint will prevent the formation of an effective seal. It is always best to renew a cracked battery especially in view of the corrosion which will be caused if the acid continues to leak.
9 If the machine has remained unused for a period of time, it is advisable to remove the battery and give it a refresher charge every six weeks or so from a battery charger. If the battery is permitted to discharge completely, the plates will sulphate and render the battery useless.
20 Occasionally, check the condition of the battery connections to ensure that they are free of corrosion and forming a good contact. If corrosion has occurred, it should be cleaned away by scraping with a knife and then using emery cloth to remove the final traces. Remake the electrical connections whilst the joint is still clean, having first smeared them lightly with petroleum jelly (not grease) to prevent recurrence of the corrosion. Badly corroded connections can have a high electrical resistance and may give the impression of a complete battery failure.

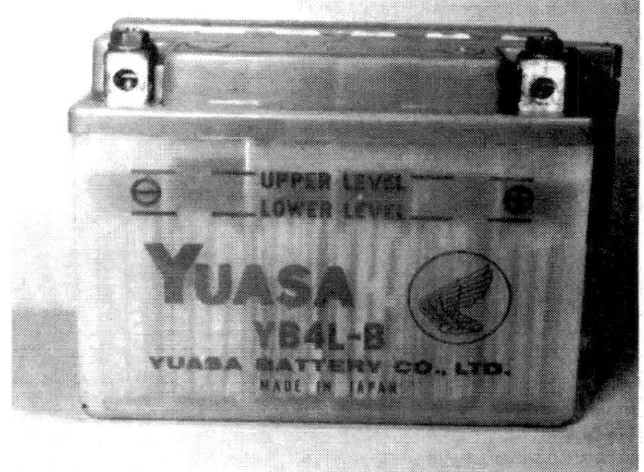

8.1a The electrolyte level can be observed through the battery casing

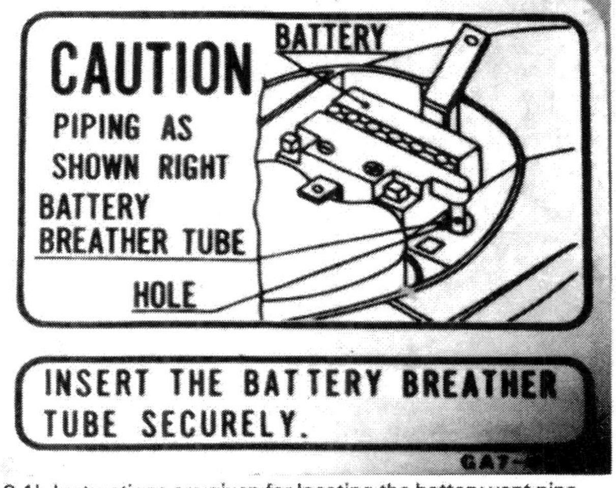

8.1b Instructions are given for locating the battery vent pipe

9 Battery: charging procedure

1 The normal charging rate for the type of battery fitted to the Honda Melody is 0.4 amp. It is permissible to charge at a more rapid rate in an emergency but this effectively shortens the life of the battery and should therefore be avoided. Because of this, go for the smallest charging rate available. Avoid 'quick charge' services offered by garages. This will indeed charge the battery rapdily but it will also overheat it and may halve its life expectancy.
2 Never omit to remove the battery cell caps or neglect to check that the side vent is clear before recharging a battery, otherwise the gas created within the battery when charging takes place might burst the case with disastrous consequences. Do not attempt to charge the battery with it in situ and with the leads still connected. This can lead to failure of the rectifier unit.
3 Ensure that the battery charger connections are correct; red to positive (+) and black to negative (–). When the battery is reconnected to the machine, the green lead must be connected to the negative terminal and the red lead to positive. This is most important, as the machine has a negative earth system. If the terminals are inadvertently reversed, the electrical system will be damaged permanently.
4 A word of caution concerning batteries. Sulphuric acid is extremely corrosive and must be handled with great respect. Do not forget that the outside of the battery is likely to retain traces of acid from previous spills, and the hands should always be washed promptly after checking the battery. Remember too that battery acid will quickly destroy clothing.
5 Note the following rules concerning battery maintenance:

Do not allow smoking or naked flames near batteries.
Do avoid acid contact with skin, eyes and clothing.
Do keep battery electrolyte level maintained.
Do avoid over-high charge rates.
Do avoid leaving the battery discharged.
Do avoid freezing.
Do use only distilled or demineralised water for topping up.

10 Resistor: location and testing

1 A resistor is fitted to the Honda Melody for the purpose of absorbing alternating current (ac) being supplied by the flywheel generator to the lighting switch when this current is not required for illuminating the headlamp, tail lamp, pilot lamp and speedometer light. The resistor is located inboard of the right-hand base support for the footrest panel. In order to test the resistor, it will be necessary to gain full access to it by removing the footrest panel; full details for doing this are contained in Chapter 4 of this Manual.
2 With the resistor thus exposed, trace the single wire running from the switch to its nearest push connector and pull apart the two halves of the connector. Set a multimeter to its resistance function and connect one of its probes to the resistor wire whilst holding the other against a good earth point on the frame. The resistance shown on the meter scale should be 6.7 ohm if the resistor is serviceable. If any other reading is obtained, then the resistor must be detached from the frame by unscrewing its single retaining bolt and replaced with a new item. Note that a poor earth connection at the resistor mounting bolt will affect the resistor performance and cause misleading resistance readings. Make sure that the earth contact is good.

11 Starter motor: removal, examination and refitting – NS50 MS models only

1 A push button switch is incorporated in the switch assembly located adjacent to the throttle twistgrip. When depressed, this switch operates a solenoid which in turn causes the starter motor to operate. The movement of the starter motor

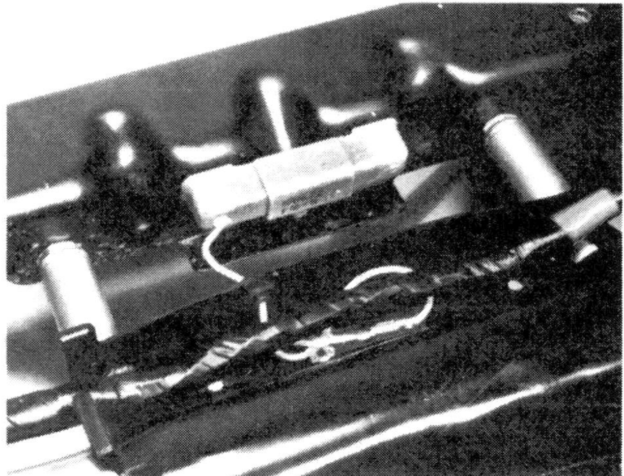

10.1 The resistor is mounted beneath the footrest panel

shaft is then passed on to the crankshaft via a pinion assembly attached to the motor itself, the teeth of which mesh with a pinion attached to the left-hand crankshaft end.
2 Commence removal of the starter motor by referring to Section 16 of Chapter 2 and removing the centre body panel from the machine, together with the complete exhaust system.
3 Move to the left-hand side of the machine and locate the electrical wires which are routed down the frame structure, just forward of the cylinder head. Pull apart the two halves of the only block connector, to disconnect the starter motor from the main wiring harness. Remove the two bolts which retain the starter motor to the transmission casing.
4 It is now necessary to detach the engine from the frame and ease it far enough rearwards to allow the starter motor to be withdrawn from its location in the transmission casing and be pulled clear of the machine. Before attempting to detach the engine, obtain a block of wood, or similar support, which will fit between the crankcase and floor, thus preventing the engine unit from dropping away from the frame directly it is detached.
5 Move to the rear wheel and unscrew the adjuster nut from the threaded end of the rear brake operating cable. Remove the cable anchor plate by unscrewing its single retaining bolt from the rear of the transmission casing. Check that the brake cable is free to slide through its guide at the front of the transmission casing. If the cable is clamped securely by the guide, then remove the guide retaining bolt.
6 Locate the single flange nut which retains the rear mudguard to the underside of the rear body panel. This nut is located in the centre of the rear edge of the mudguard. With the nut removed, the mudguard can be eased down and rearwards to free it from its forward locating tabs.
7 Unscrew the locknut from the end of the engine unit pivot bolt and remove it together with the plain washer. Using a soft-metal drift and a hammer, carefully drift the pivot bolt from position. With the bolt thus removed, carefully ease the engine unit far enough rearwards to allow full access to the oil pump retaining bolt. Rest the engine crankcase on the block positioned beneath it.
8 Release the two bolts which secure the starter motor in position. Remove the starter motor by grasping it firmly by its body and then pulling it from position. It may be found that the motor has become quite firmly stuck in the transmission casing, in which case the flange of the motor body should be gently tapped with a soft-faced hammer to help free it. Do not attempt to lever the motor away from the casing as this will only cause damage to the mating surfaces.
9 The parts of the starter motor most likely to require attention are the brushes. In order to gain access to these brushes, it is necessary to detach the pinion cover from the

body of the motor, followed by the pinion assembly and the motor body end cover. Note that it will be necessary to detach the wiring harness from the motor before the pinion cover can be removed; this being held in position by two crosshead screws with spring washers. Remove the screws which retain the pinion cover in position and pull away the cover followed by its gasket, the pinion assembly with its thrust washer, and any shims that are fitted, and the motor end cover which will be seen to contain the two brushes. The motor armature should have remained held in the motor body by the magnetic force of the body magnets; it should now be pulled out of its location in the body and placed on a clean work surface ready for inspection.

10 Both brushes will have been displaced from their holders upon separation of the end plate from the commutator upon which they bear. If either brush is worn to a length of less than 3.0 mm (0.12 in) then the brushes must be renewed as a pair. If either of the two springs contained within the brush holders is seen to be either broken or fatigued, then they too should be renewed as a pair.

11 Ensure that the commutator upon which the brushes bear is clean. If necessary, the commutator segments may be burnished using crokus paper. This is a fine abrasive paper produced for this purpose, and can be obtained from auto-electrical specialists. On no account use emery paper as it will damage the commutator. After cleaning, wipe the commutator with a rag soaked in methylated spirits or de-natured alcohol to remove any dust or grease.

12 The insulation between the individual segments of the commutator should lie a minimum distance of 2 mm (0.08 in) below the surface. If the insulation is less than this distance below the surface, it should be undercut very carefully, using a small portion of fine hacksaw blade. It may be considered worthwhile to seek the services of an auto-electrician who will be able to carry out this delicate operation at a favourable price.

13 Closely inspect the individual bars of the commutator for signs of discolouration. Any discolouration between one pair of bars indicates earthed or open armature coils, whereas discolouration on the edge of a bar indicates a high resistance. Using a multimeter set to the resistance testing function, test for continuity between the pairs of bars and between each bar and the armature shaft. If a fault is found to exist, then return the complete motor assembly (less starter pinion and cover) to a Honda service agent who will be able to confirm whether renewal of the motor is necessary. The armature is not listed as a separate item.

14 It is now advisable to carry out a general check of the remaining component parts of the starter motor. Commence by cleaning the inside of the motor body with a clean rag soaked in methylated spirits or de-natured alcohol, paying particular attention to the surface area of the permanent magnets.

15 Check the condition of the two sealing gaskets; one of which is located between the motor body and end cover and the pinion cover. If either one of these gaskets is damaged or torn, then it must be renewed. Inspect the O-ring which forms a seal between the pinion cover and transmission casing for signs of damage or deterioration. It is advisable to renew this O-ring whenever the motor is serviced.

16 Clean and carefully inspect the bearing surfaces of the armature and pinion assembly. If excessive play is felt between any two surfaces, or if the surfaces are seen to be badly scored or damaged, then the items concerned should be returned to a Honda service agent for further inspection and, if necessary, renewal. Check also the condition of the thrust washer, and any shims which are fitted, located between the pinion assembly and the motor end cover.

17 Inspect the teeth of the pinion assembly and the teeth of the pinion at the armature end. Any broken or badly worn teeth will necessitate renewal of the component concerned. Check that the small bearing located on the toothed end of the armature is free of any roughness when rotated. If this is not the case then it must be renewed. Note that Honda do not list this bearing as a separate item, any new bearing will have to be obtained from a motor factor or any of the suppliers of bearings who advertise in the Yellow Pages or motoring trade magazines.

18 On completion of carrying out the above listed examination procedure, lay out the component parts of the motor on a clean area of work surface in the order of reassembly. Commence reassembly of the motor by lightly greasing the bearing surface inside the motor body with a high melting point lithium based grease. Fit each brush and spring in their holders and check that each brush is free to slide quite freely. Ensure that used brushes are refitted in their original positions because they will have worn to the profile of the commutator. Hold both brushes fully back in their holders and carefully insert the commutator between them. Allow the brushes to be returned against the commutator and check that each one is making a proper contact. Grasp the pinion at the armature end to prevent the armature from being pulled through the end plate and then carefully insert the armature into the motor body until its end is located fully home in the bearing of the motor body. Check that the sealing gasket is correctly aligned before finally pushing the end plate against the body.

19 Lightly grease both bearing surfaces of the starter pinion assembly with a high melting point lithium based grease. Locate the thrust washer, and any shims, on the motor end of the pinion assembly and insert the assembly into the motor end plate. Apply a light smear of grease to the teeth of the pinion assembly and where a thrust washer is fitted to the cover end of the pinion assembly, slide the washer into position. Refit the sealing gasket to the motor end plate and with the gasket properly aligned, push the pinion cover into position. Refit the wiring harness and insert and tighten the assembly securing screws. Check that the O-ring is correctly seated in its groove in the pinion cover before lightly lubricating it with clean engine oil.

20 The starter motor can now be refitted to the machine by reversing the removal procedure, whilst taking care to note the following points. Ensure that the motor is pushed fully home against the transmission casing before fitting and tightening its two securing bolts. Tighten the locknut of the engine pivot bolt to a torque loading of 25 – 33 lbf ft (3.5 – 4.5 kgf m). Check that the rear brake operating cable is routed correctly and properly located in both its guide clamp and its anchor plate before the clamp and plate retaining bolts are fully tightened. With the cable end passed through the trunnion of the brake cam operating lever, fit the adjuster nut to the cable end and tighten the nut until the amount of free play measured at the tip of the handlebar lever is 10 – 15 mm (0.4 – 0.6 in). Refer to Section 6 of Chapter 2 and refit the exhaust system. Check that the electrical connection to the starter motor is correctly made before refitting the centre body panel and rear mudguard.

11.10 Check each brush for wear ...

11.11 ... and inspect the commutator assembly

11.14 Examine the surface area of the permanent magnets

11.17 Inspect the teeth of the pinion assembly

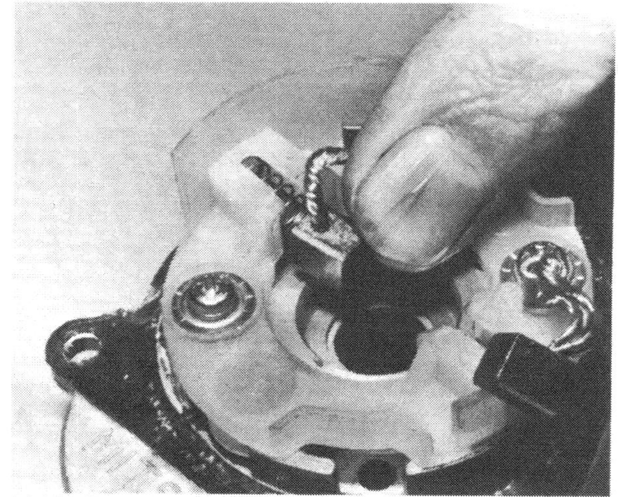
11.18a Fit each brush with spring ...

11.18b ... and carefully insert the commutator between them

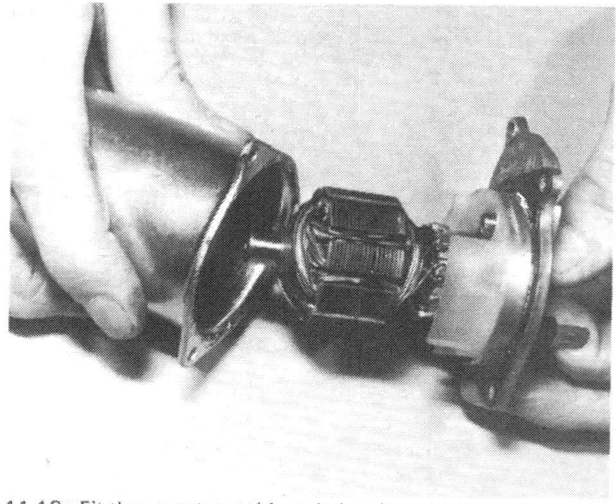
11.18c Fit the armature with end plate into the motor body

11.19a Fit the thrust washer over the shaft of the pinion assembly ...

11.19b ... and insert the assembly into the end plate

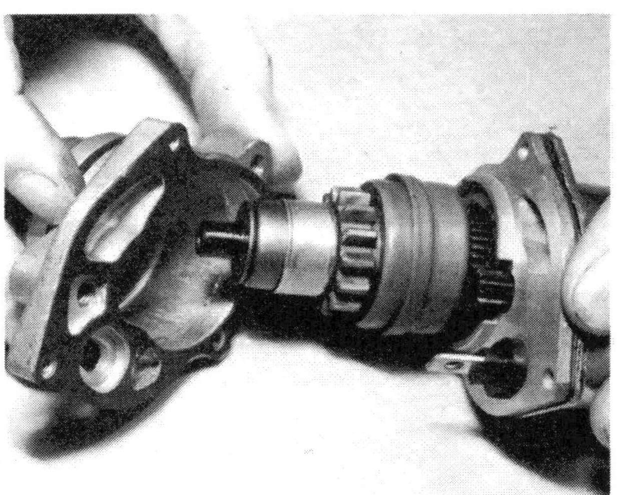
11.19c Push the pinion cover and gasket into position

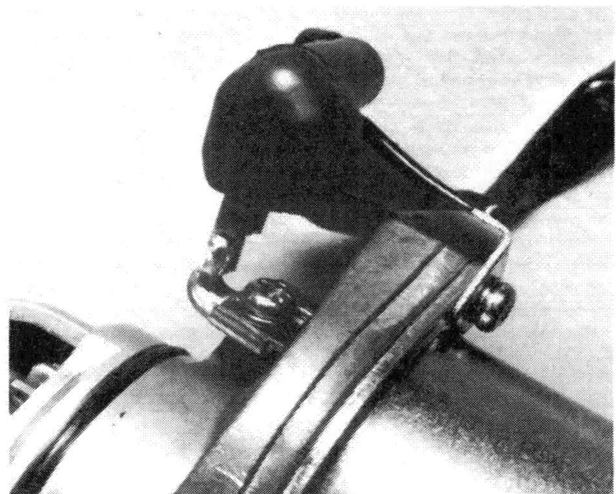
11.19d Position the wiring harness correctly

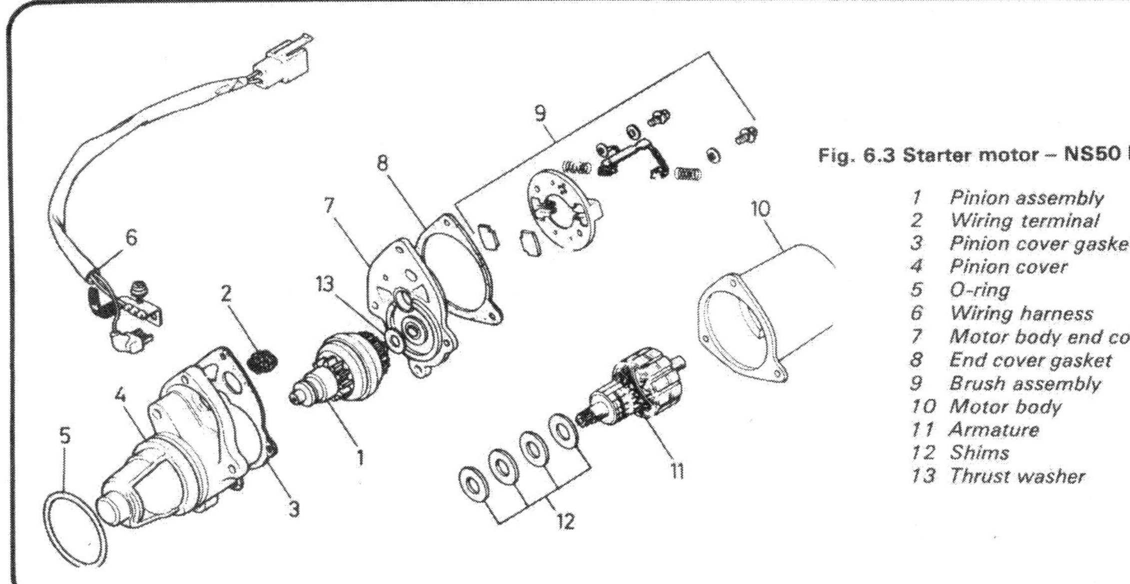

Fig. 6.3 Starter motor – NS50 MS model

1 Pinion assembly
2 Wiring terminal
3 Pinion cover gasket
4 Pinion cover
5 O-ring
6 Wiring harness
7 Motor body end cover
8 End cover gasket
9 Brush assembly
10 Motor body
11 Armature
12 Shims
13 Thrust washer

12 Starter solenoid switch: function, location and testing

1 The starter motor switch is designed to work on the electromagnetic principle. When the starter motor button is depressed, current from the battery passes through windings in the switch solenoid and generates an electro-magnetic force which causes a set of contact points to close. Immediately the points close, the starter motor is energised and a very heavy current is drawn from the battery.

2 This arrangement is used for two reasons. Firstly, the current drawn by the starter motor is very high which requires the use of proportionately heavy cables to supply current from the battery to the motor. Running such heavy cables directly to the conveniently placed handlebar start switch would be cumbersome and impractical. Second, because the demands of the starter motor are so high, as short a cable run as possible is used to minimise volt drop in the circuit. If the starter will not operate, first suspect a discharged battery. This can be checked by trying the horn or switching on the lights.

3 If the above check shows the battery to be in good condition, then suspect the solenoid switch which should come into action with a pronounced click. The switch is located on the right-hand side of the machine, just below the fuel tank and directly in front of the flasher unit. Access to the switch can be gained by removal of the centre body panel from the machine; this panel being held in position by a dome head nut with washer and a single screw. Before condemning the switch, leave it in its mounting and carry out the following test.

4 Unplug the block connector from the base of the switch and set the multimeter to its resistance function. It is now necessary to connect the terminals of a 12 volt battery to the Green/yellow and the Yellow/red wire terminals of the solenoid switch. This can be done by making up a pair of extension leads, each with a small crocodile clip attached to its end, with which to connect the battery of the machine to the switch. It is important to ensure that the positive (+) terminal of the battery is connected to the Green/yellow wire terminal.

5 With the battery thus connected, place one probe of the multimeter on the Red wire terminal of the switch and the other probe on the Red/white wire terminal. If the switch is serviceable, there should be continuity between the two terminals. If the test shows non-continuity, then the switch should be replaced with a serviceable item. Removal of the solenoid switch is accomplished simply by gripping the rubber mounting sleeve between the fingers of one hand and then pulling the switch from position.

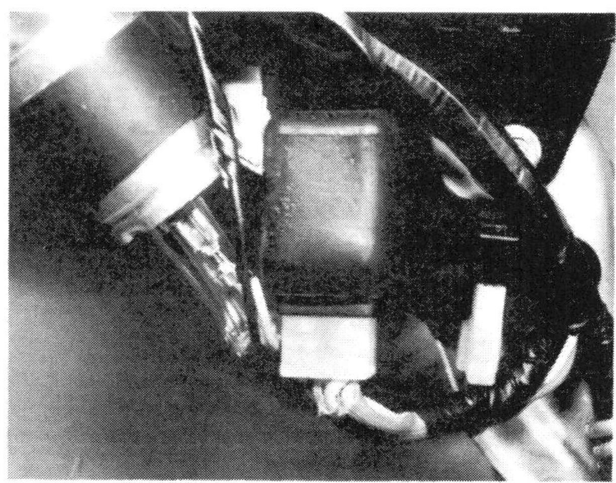

12.3 The starter solenoid switch is located between the flasher unit and silicon rectifier

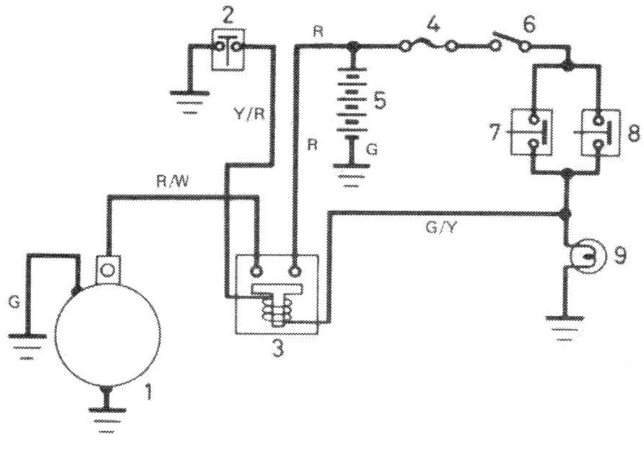

Fig. 6.4 Starter motor circuit diagram

1 *Starter motor*
2 *Handlebar switch*
3 *Starter solenoid*
4 *7 amp fuse*
5 *Battery*
6 *Ignition switch*
7 *Front brake stop lamp switch*
8 *Rear brake stop lamp switch*
9 *Stop lamp*

R – Red
Y – Yellow
W – White
G – Green

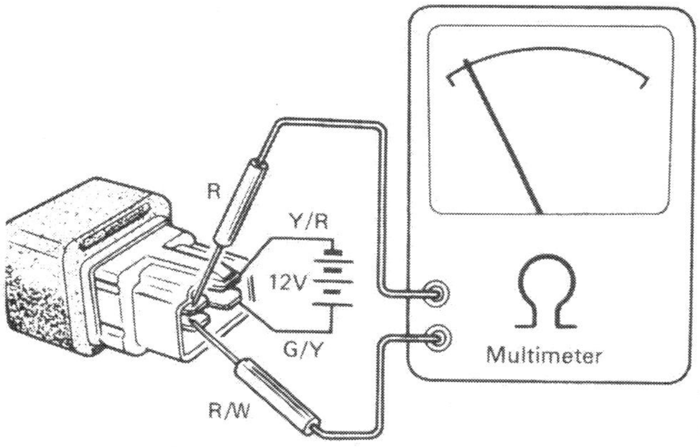

Fig. 6.5 Starter solenoid switch test

R – Red
R/W – Red and white
Y/R – Yellow and red
G/Y – Green and yellow

13 Ignition switch: location and testing

1 The ignition switch and steering lock are combined in one switch unit which is mounted on a plate which forms part of the steering head assembly. In order to gain full access to this switch, it is necessary to remove the carrier pocket from the rear of the front body panel, followed by the rear section of the front body panel. Note that only the Melody deluxe model (the NS50 MS) has the carrier pocket fitted as standard and that full instructions for the removal and fitting of body panels and cycle components may be found in Chapter 4 of this Manual. The switch unit is secured to its mounting plate by two countersunk cross head screws. Removal of the innermost of these two screws will necessitate removal of the handlebar assembly.

2 To test the ignition switch, trace the Red, Black, Green and Black/white wires that run from the switch to their block connector. It is likely that this connector has been positioned in such a way that gaining access to it is impossible without first removing the luggage basket from the front body panel, followed by the panel itself. Pull apart the two halves of the connector and turn the switch to the 'On' position. Check for continuity between the Red and the Black wire terminals of the connector.

3 Turn the switch to the 'Off' position and check for continuity between the Green and the Black/white wire terminals of the connector. Finally, carry out a check for continuity between the same two wire terminals with the switch turned to the Lock position.

4 If continuity is found to exist in all of the above tests, then the switch is serviceable. If any one of the tests shows non-continuity, then the switch must be renewed. Check, before attempting switch removal, that none of the switch wires have been cut or frayed by being in contact with the front panel mounting bracket.

14 Front and rear brake stop lamp switches: location and testing

1 The front and rear brake stop lamp switches fitted to the Honda Melody are identical units, each one being fitted in the handlebar mounting for the appropriate brake lever.

2 Each switch may be tested by first tracing the Black and the Green/yellow wires that run from the switch to their nearest push connectors. This will entail removal of the carrier pocket from the rear of the front body panel, followed by the rear section of the front body panel. In the event of the connectors still not being readily accessible, it will also entail removal of the luggage basket, followed by the front panel itself.

3 Pull apart the two halves of each connector and check for continuity between the two wires with the brake lever fully applied. If continuity is found to exist, then the switch is serviceable. If the test shows non-continuity, then the wires running to the switch should be checked for signs of chafing where they pass close to any point of the frame structure, before removing and rejecting the switch. Each switch is a straight push fit into its mounting. It is, however, important to note the small keyed location at the front of each switch during removal and fitting.

15 Fuel gauge and float switch : testing

1 In order to provide a quick and easy check on the level of fuel contained in the tank of the machine, a fuel gauge is incorporated in the instrument console assembly. This gauge is operated by a float switch. If the gauge is thought to be malfunctioning then both it and the float switch should be tested by using the following procedure.

14.1 Each stop lamp switch is incorporated within its respective brake lever assembly (arrowed)

2 Commence by raising the seat of the machine to expose the top of the fuel tank and removing the centre body panel to partially expose the sides of the tank. This panel is retained in position with a dome head nut with washer and a single screw. Trace the two electrical wires which run from the float switch to their nearest push connectors and pull apart the two halves of the connectors.

3 Turn the ignition switch to the 'On' position and operate the direction indicator lamps to ensure that the battery circuit is functioning correctly. Switch off the indicator lamps but leave the ignition on. Select the two wire connections on the gauge side of the push connectors and touch them together. The needle on the fuel gauge should deflect to its 'full' position. If this is not the case, then there is a fault in the wiring to the gauge or in the gauge itself. Refer to Section 3 of this Chapter and carry out a visual check on the wiring. Alternatively, carry out a check for continuity of the wiring as described in Section 2. Details of dismantling the instrument console in order to gain access to the fuel gauge are given in Section 12 of Chapter 4.

4 If the fuel gauge and its associated wiring are found to be serviceable, then the float switch itself must be suspect. This switch can be removed from the fuel tank for testing by turning its retaining ring anti-clockwise to release it from the spigots on the tank housing. Honda recommend that special tool No 07920-GA70000 is used for this purpose but it was found that inserting the nose ends of a pair of long-nose pliers into the ring slots and then turning the pliers served to rotate the ring. With the ring thus removed, the switch can be drawn out of the tank, whilst taking great care to avoid bending the float arm.

5 Set a multimeter to its resistance function and connect one of its probes to each lead of the switch. Measure the resistance with the float moved to both ends of its range. If the readings obtained differ from those given below, then the switch is unserviceable and should be renewed.

Switch position	Resistance reading
Upper (tank full)	4 – 10 ohms
Lower (tank empty)	90 – 100 ohms

6 Finally, before fitting the switch in the tank, carry out a final check on operation by reconnecting the wiring connections and then moving the float over its full range whilst observing the position of the gauge needle. With the float moved to its fully up position, the gauge needle should be seen to move to its 'Full' position. As the float is lowered, the needle should be seen to move across the gauge until it reaches its 'Empty' position. On completion of this check, fit the switch into the tank. Ensure that the arrow marked on the retaining ring aligns with the arrow on the tank housing once it is fully tightened.

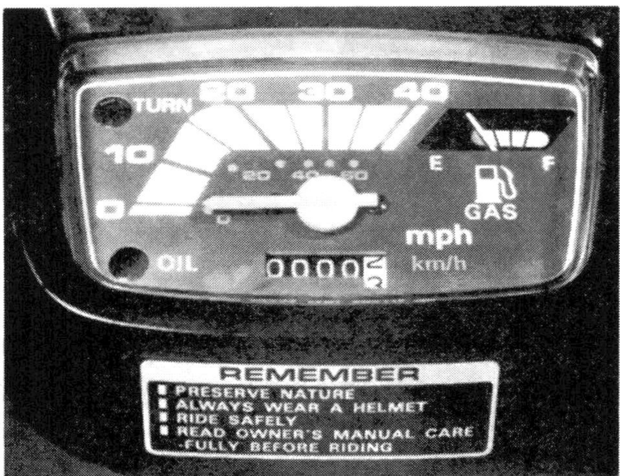

15.1 A fuel gauge is incorporated in the instrument console assembly

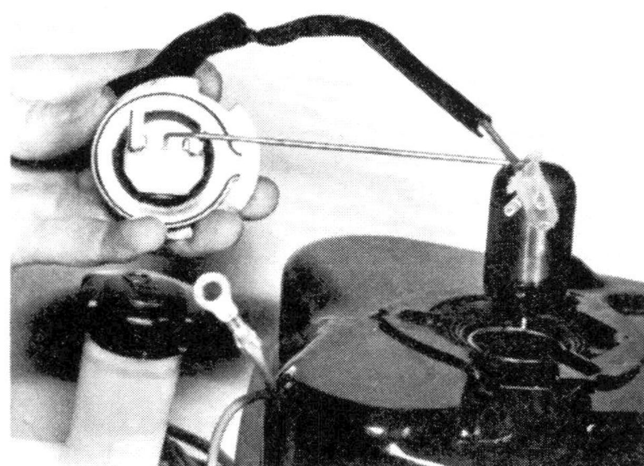
15.4 The float switch can be drawn out of the fuel tank

15.6 Align the float switch retaining ring

16 Low oil level warning lamp circuit: testing

1 The circuit consists of a float-type switch mounted in the top of the oil tank and the bulb itself, which is mounted in the speedometer.

2 When the ignition is switched on the lamp should light for approximately 5 seconds as a check that the system is working, then it will go out. If it does not go out there is insufficient (200 cc/0.35 pint or less) oil in the tank; top up the tank as described in Routine Maintenance and check again that the lamp goes out. If the lamp does not light at all the system is faulty and must be checked.

3 The most likely cause of failure will be a blown bulb, but it is worth checking first that the battery is fully charged; check that all other systems are working normally as a quick test of this. The bulb is removed and refitted as described in Section 26 of this Chapter.

4 If the bulb is in good condition the fault must lie in the switch or in the wiring. Remove the rear body panel to gain access to the oil tank top surface and disconnect the switch at the multi-pin block connector joining it to the main loom. Test the wiring in the main loom as follows.

5 First check that full battery voltage is available, connecting a dc voltmeter between the black wire terminal and a suitable earth point on the frame and switching on the ignition. If battery voltage is not measured check the black wire back to the ignition switch and the switch itself. Next use a multimeter set to one of the resistance scales to check that there is continuity between the green wire terminal and a good earth point on the frame, and between the green/red wire terminal and the warning lamp bulb holder; if resistance is measured in either case the wire is damaged or broken and must be checked carefully until the fault is found.

6 If the bulb and wiring are in good condition the fault must lie in the switch. Unplug it from the oil tank and wash off all traces of oil, checking that the terminals are clean and making good contact. Use a multimeter set to the appropriate resistance scale to test the switch. With the float at the bottom of the switch column, measure the resistance between the switch green/red and black wire terminals; a reading of 5 – 15 ohms should be obtained. Check that there is no continuity (ie infinite resistance) between the green and black wire terminals when the float is fuly lowered. Finally lift the float fully to the top of the column and measure again the resistance between the green/red and black wire terminals with the switch connected to the main loom and the ignition switched on; the reading should now be 340 ohms.

7 If any of the above tests reveals a fault in the switch, or if obvious signs of damage can be seen on removing the switch from the oil tank, the switch must be renewed; repairs are not possible.

8 To test the accuracy of the switch, remove it from the tank and clean it, as described above, then connect it again to the main loom, switch on the ignition, raise the float to the top of the column and wait for the lamp to go out. When the lamp goes out move the float to the bottom of the column, whereupon the lamp should light again, and slowly move the float back up the column until the lamp goes out. Measure the distance between the float stop at the bottom of the column and the bottom of the float; the distance should be 4.5 mm (0.18 in) although a tolerance of 1.0 mm (0.04 in) above or below this figure is allowed. If the lamp lights above or below these limits the switch is inaccurate.

9 The only solution to an inaccurate switch is to renew it, although it must be up to the owner to decide whether this is justified; it is a simple matter to drain the oil tank and to pour in oil until the lamp goes out, measuring carefully and recording for future reference the amount of oil necessary to do this.

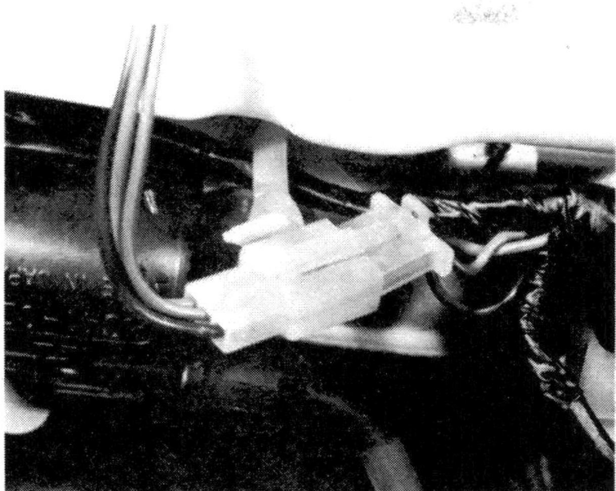
16.4 Disconnect the oil level float switch block connector

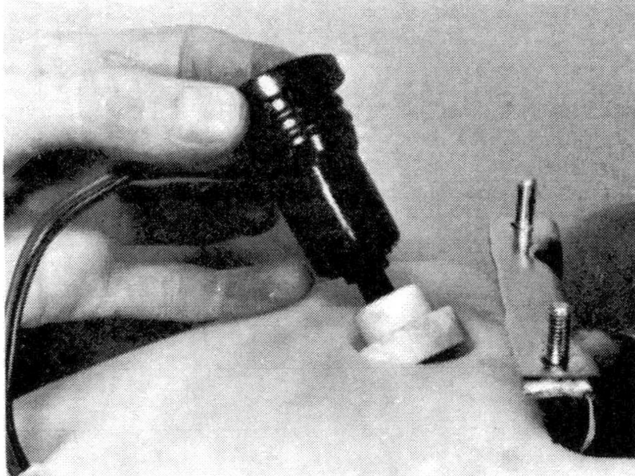
16.6 The float switch can be unplugged from the oil tank

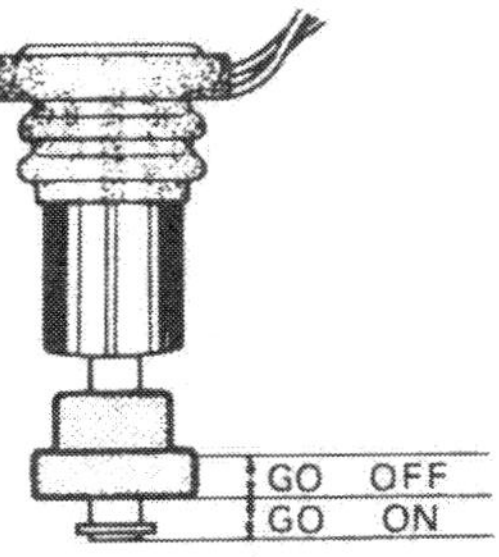

Fig. 6.6 Oil level warning light switch float measurement

4.5 ± 1.0 mm

17 Handlebar switches: general information

1 Generally speaking, the switches give little trouble, but if necessary, they can be dismantled by separating the halves which form a split clamp around the handlebars.

2 Always disconnect the battery before removing any of the switches, to prevent the possibility of a short circuit. Most troubles are caused by dirty contacts, but in the event of the breakage of some internal part, it will be necessary to renew the complete switch.

3 Because the internal components of each switch are very small, and therefore difficult to dismantle and reassemble, it is suggested a special electrial contact cleaner be used to clean corroded contacts. This can be sprayed into each switch, without the need for dismantling.

4 It will be necessary to obtain a multimeter and set it to its resistance function in order to carry out the various continuity checks described in the following Sections.

18 Starter switch: location and testing – NS50 MS models only

1 The starter switch is fitted to Honda Melody models equipped with a starter motor and takes the form of a push button which is incorporated in the switch assembly located adjacent to the throttle twistgrip.

2 To test the switch, remove the headlamp unit from its location in the handlebar nacelle and trace the Yellow/red wire running from the switch to its nearest push connector. Pull apart the two halves of this connector. Connect one probe of the multimeter to the wire connection and the other probe to a good earth point (the head of one of the switch securing screws is ideal). Press the starter button and check for continuity as the switch contacts come together to form a closed circuit. If continuity is found, then the switch is serviceable. If the test shows non-continuity, then the complete switch assembly must be renewed; this is assuming that there is no defect in the wire leading to the switch or that the switch contacts are free from contamination.

19 Lighting switch: location and testing

1 The lighting switch is incorporated in the upper half of the switch assembly located adjacent to the throttle twistgrip.

2 To test the switch, remove the headlamp unit from the handlebar nacelle and trace the Pink, Yellow, Brown and Brown/white wires running from the switch to their nearest push connectors. Pull apart these connectors and move the switch lever to the position marked with a dot; that is, nearest to the twistgrip rubber. Check for continuity between the Pink and the Yellow wire terminals.

3 Move the switch lever to the position marked with a P and check for continuity between the Pink and Yellow wire terminals, the Pink and Brown/white wire terminals and the Yellow and Brown/white wire terminals. Finally, move the switch lever to the position marked with the letters HL and check for continuity between the Yellow and Brown wire terminals, the Yellow and Brown/white wire terminals and the Brown and Brown/white wire terminals.

4 If continuity is found to exist in all of the above tests, then the switch is serviceable. If any one of the tests shows non-continuity, then the switch must be renewed as any failure in the switch can plunge the lighting system into darkness with disastrous consequences.

20 Direction indicator switch: location and testing

1 The direction indicator switch is incorporated in the switch assembly located adjacent to the left-hand handlebar grip rubber. The switch may be tested by removing the headlamp unit from its location in the handlebar nacelle and then tracing the Grey, Light blue and Orange wires that run from the switch

to their nearest push connectors. Pull apart these connectors and move the switch lever to the left. Check for continuity between the Grey and the Orange wire terminals.
2 Move the switch lever fully right and check for continuity between the Grey and the Light blue wire terminals. If continuity is found to exist in both of these tests, then the switch is serviceable. If either one of the tests shows non-continuity, then the switch must be renewed; this is assuming that there is no defect in the wires leading to the switch or that the switch contacts are free from contamination.

21 Headlamp dipswitch: location and testing

1 The headlamp dipswitch is incorporated in the upper half of the switch assembly located adjacent to the left-hand handlebar grip rubber.
2 To test the switch, remove the headlamp unit from its location in the handlebar nacelle and trace the Brown, Blue and White wires running from the switch to their nearest push connectors. Pull apart these connectors and set the switch button to its main beam position. Check for continuity between the Brown and the Blue wire terminals.
3 Move the switch button to its dipped beam position and check for continuity between the Brown and White wire terminals. Finally, move the switch button to a point which is midway between the above mentioned positions and check for continuity between the Brown and Blue wire terminals, the Brown and White wire terminals and the Blue and White wire terminals.
4 If continuity is found to exist in all of the above tests, then the switch is serviceable. If any one of these tests shows non-continuity, then the switch must be renewed, as any failure that occurs when changing from one beam to the other can plunge the lighting system into darkness, with disastrous consequences.

22 Horn switch: location and testing

1 The horn switch takes the form of a push button which is incorporated in the lower half of the switch assembly located adjacent to the left-hand handlebar grip rubber.
2 To test the switch, remove the headlamp unit from its location in the handlebar nacelle and trace the Light blue and Green wires running from the switch to their nearest push connectors. Pull apart these connections with the push button pressed fully in. If continuity is found, then the switch is serviceable. If the test shows non-continuity, then the complete switch assembly must be renewed; this is assuming that there is no defect in the wires leading to the switch or that the switch contacts are free from contamination.

23 Horn: location, adjustment and replacement

1 The horn is located at the base of the frame steering head, just above the mounting bracket for the front body panel, and may be viewed through a grille which forms part of the panel moulding. Access to the horn can be gained after removal of the carrier packet from the rear of the front body panel followed by the luggage basket and the panel itself. Note that only the Melody deluxe model (the NS50 MS) has the carrier pocket fitted as standard. Full instructions for the removal and fitting of body panels and cycle parts may be found in the appropriate Section of Chapter 4.
2 The horn should not normally require adjustment. If, however, this becomes necessary, then provision for adjustment is provided by means of a screw at the rear of the horn case. This screw should be turned fractionally clockwise or anti-clockwise to vary the horn volume. Note the locknut at the base of the screw; this should be loosened to facilitate rotation of the screw and then tightened on completion of adjustment.
3 If necessary, the horn may be removed from the machine by disconnecting the two electrical wires from the horn terminals, after having first noted their fitted positions, and then unscrewing the single bolt which serves to retain the horn mounting plate to the body panel mounting bracket. Fitting a replacement horn is a direct reversal of the removal procedure.

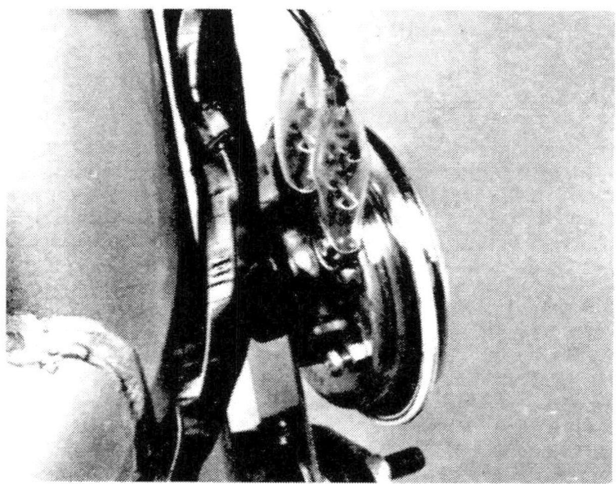

23.1 The horn is located at the base of the frame steering head

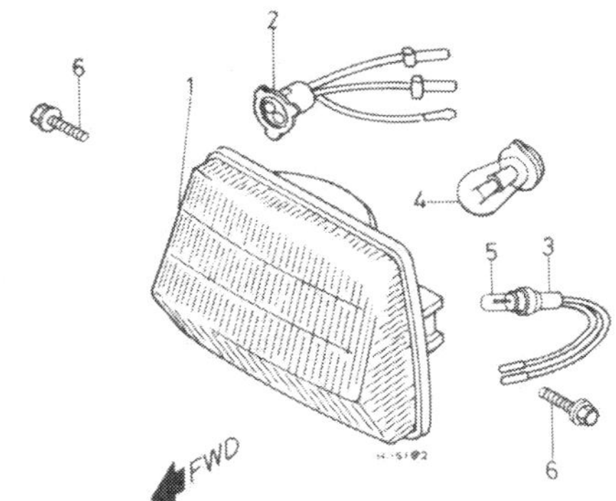

Fig. 6.7 Headlamp

1 Headlamp assembly
2 Bulb holder
3 Parking lamp bulb holder
4 Headlamp bulb
5 Pilot lamp bulb
6 Bolt – 2 off

24 Headlamp: bulb renewal and beam alignment

1 In order to gain access to the headlamp bulbs, it is necessary to remove the headlamp lens and reflector assembly from the moulded plastic nacelle which is fitted over the handlebars. The headlamp unit may be removed by unscrewing each of the two retaining bolts which pass through the sides of the nacelle and into the reflector moulding. With these bolts removed, the headlamp unit can be eased forward out of the nacelle.
2 The main headlamp bulb is a push fit in the central bulb location of the reflector and is retained in position by a holder which is fitted into the location over it. The bulb can be fitted in one position only to ensure that it is always correctly focussed. A twin filament bulb of a 12 volt, 25/25 watt rating is fitted as standard. To remove the bulb, push in on the holder and turn it

anti-clockwise before pulling it clear of the reflector. Grasp the end of the bulb and carefully pull it out of its location.
3 The holder for the pilot lamp bulb is a direct push fit into the reflector. The bulb is of a 12 volt, 3.4 watt rating and can be removed from its holder by pushing it inwards, turning it anti-clockwise and then pulling it outwards.
4 Beam height adjustment is effected by unscrewing each of the headlamp unit retaining bolts just enough to allow the unit to be tilted in relation to the handlebar nacelle. No provision is made for adjusting the horizontal alignment of the beam.
5 In the UK, regulations stipulate that the headlamp must be arranged so that the light will not dazzle a person standing at a distance greater than 25 feet from the lamp, whose eye level is not less than 3 feet 6 inches above that plane. It is easy to approximate this setting by placing the machine 25 feet away from a wall, on a level road, and setting the dip beam height so that it is concentrated at the same height as the distance of the centre of the headlamp from the ground. The rider must be seated normally during this operation and also the pillion passenger, if one is carried regularly.
6 Most other areas have similar regulations controlling headlight beam alignment, and these should be checked before any adjustment is made.

25 Stop and tail lamp: bulb renewal

1 The combined stop and tail lamp unit fitted to the Honda Melody is fitted with a double filament bulb having offset pins to prevent its unintentional reversal in the bulb holder. The lamp unit serves a two-fold purpose – to illuminate the rear of the machine and the rear number plate, and to give visual warning when either brake is applied.
2 To gain access to the bulb, remove the two screws which hold the red plastic lens in position. The bulb is released by pressing inwards and with an anti-clockwise turning action before pulling it free of its holder. Be careful, when lifting the lens from its mounting plate, not to tear the sealing gasket located between the two items. This gasket is essential to keep moisture and dirt from entering the electrical contacts within the unit.
3 Refitting of the bulb and lens is a reversal of the removal procedure. Check the inside of the bulb holder for any signs of corrosion or moisture, ensuring that the contacts are free to move when depressed. When refitting the lens securing screws, take care not to overtighten them as it is possible that the lens may crack.

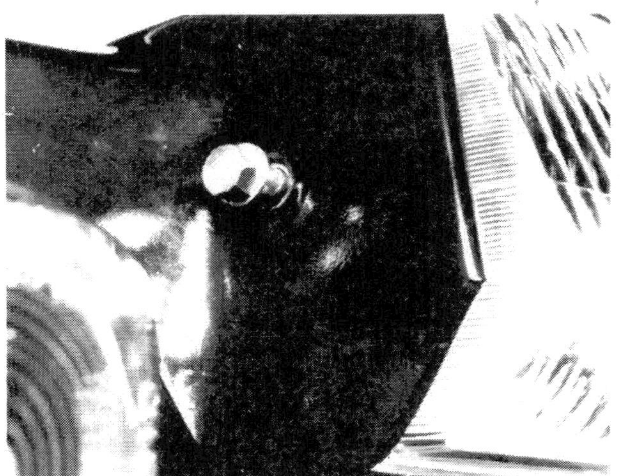

24.1 Unscrew the headlamp unit retaining bolts

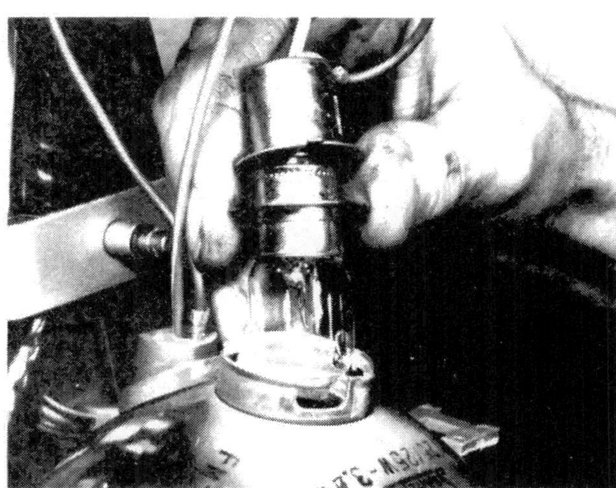

24.2 The headlamp bulb is a push fit in the centre of the reflector

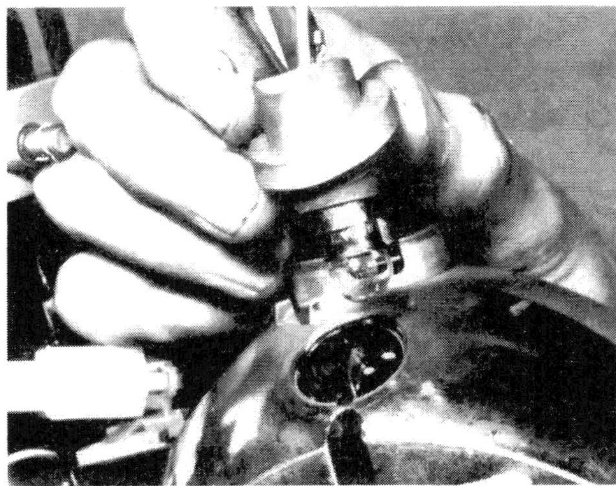

24.3 The pilot lamp bulb holder can be pulled out of its location in the reflector

25.3a Fit the stop/tail lamp bulb ...

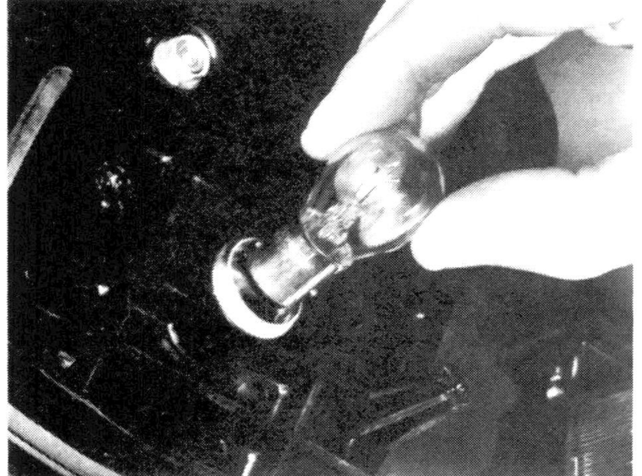
25.3b ... check the lens sealing gasket ...

25.3c ... and do not overtighten the lens securing screws

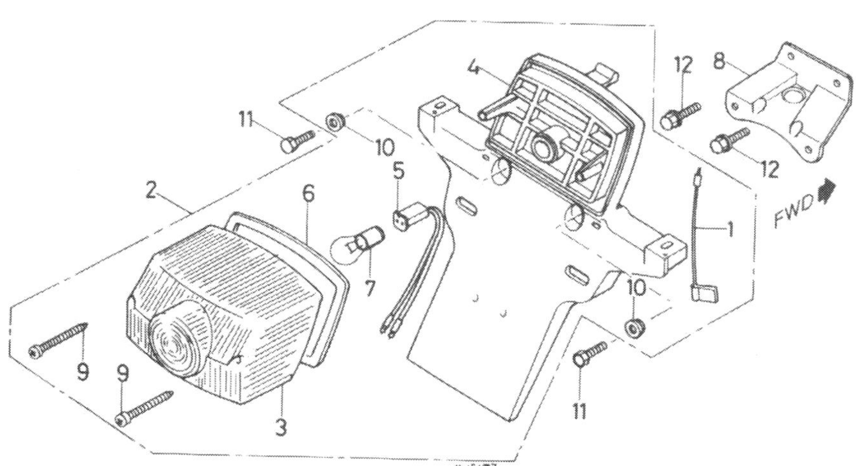

Fig. 6.8 Tail lamp

1 Earth lead
2 Tail lamp assembly
3 Lens
4 Mounting bracket
5 Bulb holder
6 Sealing gasket
7 Bulb
8 Mounting bracket
9 Screw – 2 off
10 Collar – 2 off
11 Bolt – 2 off
12 Bolt – 2 off

26 Instrument console assembly: bulb renewal

1 Access can be gained to the various warning and instrument illumination bulbs contained within the instrument console by detaching the two halves of the nacelle from the handlebars and then unplugging each bulb holder from the base of the console.

2 Commence removal of the nacelle by unscrewing and removing the single screw from the centre of the rear face of the lower half of the nacelle. Follow this by removing the two screws which pass upwards through the upper edge of the lower half of the nacelle; the lower half of the nacelle can now be detached from the handlebars to expose the base of the console. It may, at this stage, be possible to remove the required bulb holder. If this proves to be difficult, then unscrew the knurled cap which retains the speedometer cable to the base of the instrument. Remove each of the two headlamp unit retaining bolts from the upper half of the nacelle, pull the headlamp unit from position and lift the upper half of the nacelle up away from the handlebars to fully expose the base of the instrument console.

3 Each bulb holder is a straight push fit in the base of the instrument console. With the holder pulled from position, its bulb may be removed by pushing it inwards, turning it anti-clockwise and then pulling it from position.

4 Before fitting a replacement bulb, check that the contact within the holder is free to move when depressed. Check also that the inside of the holder is free of any corrosion or moisture. Fitting the bulb and holder and reassembling the nacelle around the handlebars is a direct reversal of the above listed procedure. If the upper half of the handlebar nacelle was disturbed, then do not omit to realign the beam of the headlamp before taking the machine onto the road (see Section 24 of this Chapter).

27 Flashing indicator lamp: bulb renewal

1 Flashing indicator lamps are fitted to the front and rear of the Honda Melody. Access to each bulb is gained by removing the single screw which retains each plastic lens cover in position and then unclipping the lens from its mounting whilst taking care not to tear the seal located between the two items.

2 The bulbs fitted are of the bayonet type and may be released by pushing in, turning anti-clockwise and pulling out of the holder. Each bulb is of a 12 volt, 21 watt rating.

3 Refitting of the bulb and lens is a reversal of the removal

procedure. Check the inside of the bulb holder for any signs of corrosion or moisture, ensuring that the contact is free to move when depressed. Do not omit to fit a lens seal which is in good condition and take great care not to overtighten the lens securing screw as it is possible to crack the lens by doing so.

28 Flasher unit: location and renewal

1 The flasher relay unit is located on the right-hand side of the machine, just below the fuel tank and directly to the rear of the starter relay. Access to this unit may be gained by removal of the centre body panel. This panel is held in position by a dome head nut with washer and a single screw.

2 If the flasher unit is functioning correctly, a series of audible clicks will be heard when the indicator lamps are in operation. If the unit malfunctions and all the bulbs are in working order, the usual sympton is one initial flash before the unit goes dead; it will be necessary to replace the unit complete if the fault cannot be attributed to any other cause.

3 If renewal is necessary, then removal of the existing unit is simple. Note the fitted position of each of the push connectors attached to the base of the unit and then unplug each one. Grasp the rubber mounting sleeve with the fingers of one hand and then extract the unit from the sleeve. When fitting a replacement unit, take great care not to subject it to any sudden shock as it is easily damaged if dropped.

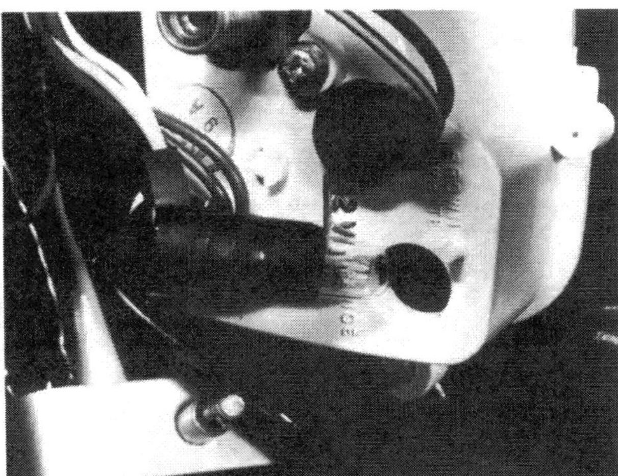

26.1 Each bulb holder is a push fit in the instrument console

27.1a Remove the indicator lens securing screw

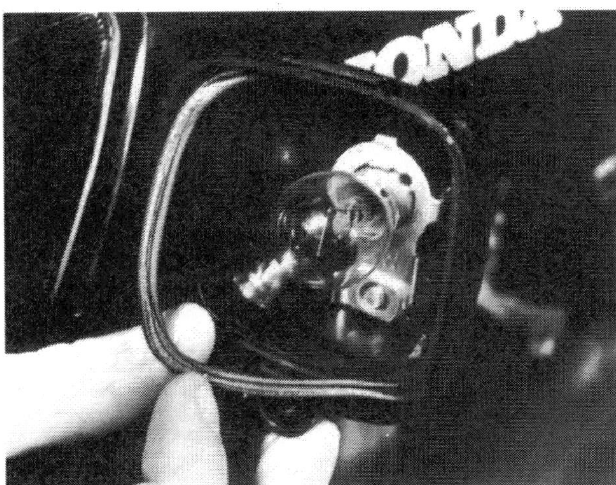

27.1b Check the lens seal for damage

27.2 Each bulb fitting is of the bayonet type

Fault diagnosis overleaf

29 Fault diagnosis : electrical system

Symptom	Cause	Remedy
Complete electrical failure	Blown fuse	Check wiring and electrical component for short circuit before fitting new fuse.
	Isolated battery	Check battery connections, also whether connections show signs of corrosion
Dim indicator lamps, horn and starter motor inoperative	Discharged battery	Remove battery and service. Check output of charging system.
Constantly blowing bulbs	Vibration or poor earth connection	Check security of bulb holders. Check earth return connections.
	Defective or poorly earthed resistor	Test and, if necessary, renew resistor.
Starter motor sluggish	Worn brushes	Remove starter motor and renew brushes.
	Dirty commutator	Clean.
Starter motor does not operate	Faulty switches or wiring	Check continuity.
	Battery flat	Recharge.
	Loose battery terminal connection(s)	Check and tighten if necessary.
Flashing indicators do not operate, or flash fast or slow	Blown bulb	Renew bulb.
	Damaged flasher unit	Renew flasher unit.

The ND50 M-C Melody II deluxe model

The NP50 D Melody mini model

Chapter 7 The NB, ND and NP50 models

Contents

Specifications

Note: Information is given only where it differs from that given in the Specifications Sections of Chapters 1 to 6. Except where otherwise specified, information applies to all models.

Model dimensions and weights

ND50:	
Overall length	1525 mm (60.0 in)
Wheelbase	1095 mm (43.1 in)
Dry weight	57 kg (126 lb)
NB50:	
Overall length	1555 mm (61.2 in)
Ground clearance	100 mm (3.9 in)
Wheelbase	1120 mm (44.1 in)
NP50:	
Overall length	1490 mm (58.7 in)
Overall width	585 mm (23.0 in)
Overall height	915 mm (36.0 in)
Ground clearance	95 mm (3.7 in)
Wheelbase	1080 mm (42.5 in)
Dry weight	38 kg (84 lb)

Specifications relating to Chapter 1

Engine

Compression ratio:	
ND50	6.8 : 1
NB and NP50	7.2 : 1

Piston and rings – NP50

Cylinder bore to piston clearance	0.030 – 0.060 mm (0.0012 – 0.0024 in)

Kickstart assembly – NB and NP50

Idle gear pinion shaft OD – at left side	As right side	

Clutch

	Standard	Service limit
Driven face spring free length:		
NB50	87.9 mm (3.46 in)	82.8 mm (3.26 in)
NP50	74.0 mm (2.91 in)	70.0 mm (2.76 in)

Transmission

Type – NP50	Automatic single speed
Primary reduction ratio:	
ND50	2.40 – 1.20 : 1
NB50	2.65 – 1.20 : 1
NP50	1.30 : 1
Final reduction ratio	8.852 : 1
Drive belt width – NP50:	
Standard	10.7 mm (0.42 in)
Service limit	9.6 mm (0.38 in)
Drive face boss OD – NB50:	
Standard	21.995 – 22.025 mm (0.8659 – 0.8671 in)
Service limit	21.960 mm (0.8646 in)
Sliding drive face bush ID – NB50:	
Standard	22.035 – 22.095 mm (0.8675 – 0.8699 in)
Service limit	22.130 mm (0.8713 in)

Torque wrench settings

Rear wheel retaining nut – NP50	47 – 58 lbf ft	6.5 – 8.0 kgf m

Specifications relating to Chapter 2

Fuel tank – NP50

Overall capacity	2.5 litre (0.55 Imp gal)

Carburettor

ID number:	
ND50	PA19B-A
NB50	PA05A
NP50	PA23A-A
Venturi diameter – NB50	14 mm (0.55 in)
Pilot screw setting – turns out from fully in:	
ND50	1 turn
NB50	$1\frac{3}{8}$ turns
NP50	$1\frac{1}{2}$ turns
Needle clip position – NB50	3rd groove from top

Lubrication system

Oil tank capacity – NB and NP50	0.8 litre (1.4 Imp pint)

Specifications relating to Chapter 3

Spark plug – ND50

Make	NGK	ND
Type	BPR6HS	W20FPR

Specifications relating to Chapter 4

Front forks

Type – NP50	Telescopic, undamped	
Spring free length:	**Standard**	**Service limit**
NB50	131.0 mm (5.16 in)	127.1 mm (5.00 in)
NP50	101.1 mm (3.98 in)	98.1 mm (3.86 in)

Rear suspension

Travel:		
ND and NB50	60 mm (2.36 in)	
NP50	50 mm (1.97 in)	
Spring free length:	**Standard**	**Service limit**
ND50	181.0 mm (7.13 in)	177.4 mm (6.98 in)
NB50	195.7 mm (7.71 in)	189.8 mm (7.47 in)
NP50	192.9 mm (7.59 in)	187.1 mm (7.37 in)

Torque wrench settings

Steering stem nut – NB50	58 – 87 lbf ft	8.0 – 12.0 kgf m
Handlebar clamp bolt – NP50	43 – 65 lbf ft	6.0 – 9.0 kgf m

Specifications relating to Chapter 5

Tyres

Size – front and rear:	
ND and NB50	2.75 x 10-4PR
NP50	2.50 x 10-2PR
Tyre pressure – tyre cold:	
Front tyre – NP50	18 psi (1.25 kg/cm²)

Torque wrench settings

Rear wheel retaining nut – NP50	47 – 58 lbf ft	6.5 – 8.0 kgf m

Specifications relating to Chapter 6

Electrical system

Voltage – NP 50	6V
Generator output – @ 5000 rpm:	
ND50	78W
NB50	80W
NP50	75W

Charging system

	Lights off (Day)	Lights on (Night)
Output – NB50:		
@ 1800 rpm or less	14.2V	–
@ 2000 rpm or less	–	14.2V
@ 2500 rpm	17.4V (0.7A) or more	17.1V (0.3A) or more
@ 6000 rpm	18.0V (5.0A) or less	17.6V (1.0A) or more
Output – NP50:		
@ 1000 rpm or less	7.0V	–
@ 1500 rpm or less	–	7.0V
@ 2500 rpm	8.5V (0.5A) or more	8.3V (0.25A) or less
@ 6000 rpm	8.8V (2.5A) or less	8.6V (2.5A) or less

Battery – NP50

Make and type	N/Av
Voltage	6V

Fuse rating – NP50 .... 10A

Bulbs – NB50

Indicator lamps	12V 15W
Oil level warning light	LED
Speedometer light	12V 1.7W

Bulbs – NP50

Headlamp	6V 18/18W
Tail/stop lamp	6V 5/21W
Pilot lamp	6V 3W
Indicator lamps	6V 17W
Oil level warning light	LED
Speedometer light	6V 3W

1 Introduction

The first six Chapters of this Manual cover the NS50 Melody and Melody deluxe models sold in the UK from 1981 to 1984; this Chapter covers the ND50 Melody II deluxe, the NB50 Melody deluxe and the NP50 Melody Mini, listing any changes in specification and any significant differences in working procedures.

When working on one of these later models, first refer to this Chapter for information; if none is given the task will be substantially the same as that described in the relevant part of Chapters 1 to 6.

The first problem when working on one of these later machines is to identify it exactly. This is usually necessary in working procedures and is essential when replacement parts are being purchased. The identifying features of the various models are given below:

ND50 M-C Melody II deluxe

Sold from April 1982 to April 1985 this model can be identified by its frame numbers which begin at AB07-2259453. Basically unchanged from the NS50 MS-B Melody deluxe, this model differs in having a fully enclosed tail section with a storage compartment on the right-hand side and revised mountings for the tail lamp and flashing indicator lamps. It is fitted as standard with a storage compartment on the inside of the front legshields and is most easily recognised by the chrome-plated covers attached to the front fork links.

NB50 MS-E Melody deluxe

Sold from March 1984 to April 1985 this machine replaced the earlier NS50 MS-B Melody deluxe. It is fitted with black plastic covers over the front fork bottom links, a storage compartment behind the front legshields and a lockable storage compartment in the right-hand side of the fully-enclosed tail section. The most obvious identifying feature for this model is its flashing indicator lamps; the front lamps are set into the handlebar nacelle and extend from the headlamp to the lever assembly on each side, while at the rear the indicator lamps form part of the tail lamp assembly.

NP50 D Melody Mini

Sold from February 1983 to April 1985 this model's frame numbers begin at AB14-1500001. It is a much less sophisticated machine, with a single-speed transmission, manual choke and fuel tap, and 6 volt electrical system. Although very similar in appearance to the earlier NS50 D Melody model it is easily distinguished by its telescopic front forks.

2 Engine/gearbox unit: modifications – NB and NP 50 models

1 With reference to the relevant Sections of Chapter 1, note the following modifications in working procedure:

NB50

2 The cylinder head cowling is now in one piece, and is retained by two bolts.

3 The cooling fan is now retained by two bolts. The rotor design has been modified which means that it is now necessary to use the correct Honda service tool to remove it. Refer to Section 3.

4 The transmission casing is retained by eight bolts.

5 The kickstart idler pinion shaft coil spring (item 14, Fig. 1.1) is no longer fitted, and the shaft itself is now of the same outside diameter at both bearing surfaces.

6 In the reduction gearbox the output shaft pinion is now separate and is located on the shaft by two circlips. Also the shaft now rotates in a ball bearing at its left-hand end.

NP50

7 The engine/gearbox unit fitted to this model has also received the modifications detailed in paragraphs 2 – 5 above.

8 The transmission has been modified; the crankshaft mounted centrifugal clutch has been replaced by a simple pulley so that only one speed is available. The rear clutch remains the same but its spring-loaded sliding face is now little more than a belt-tensioning device. Removal and refitting of the crankshaft-mounted components remains as described for the NS50 D Melody.

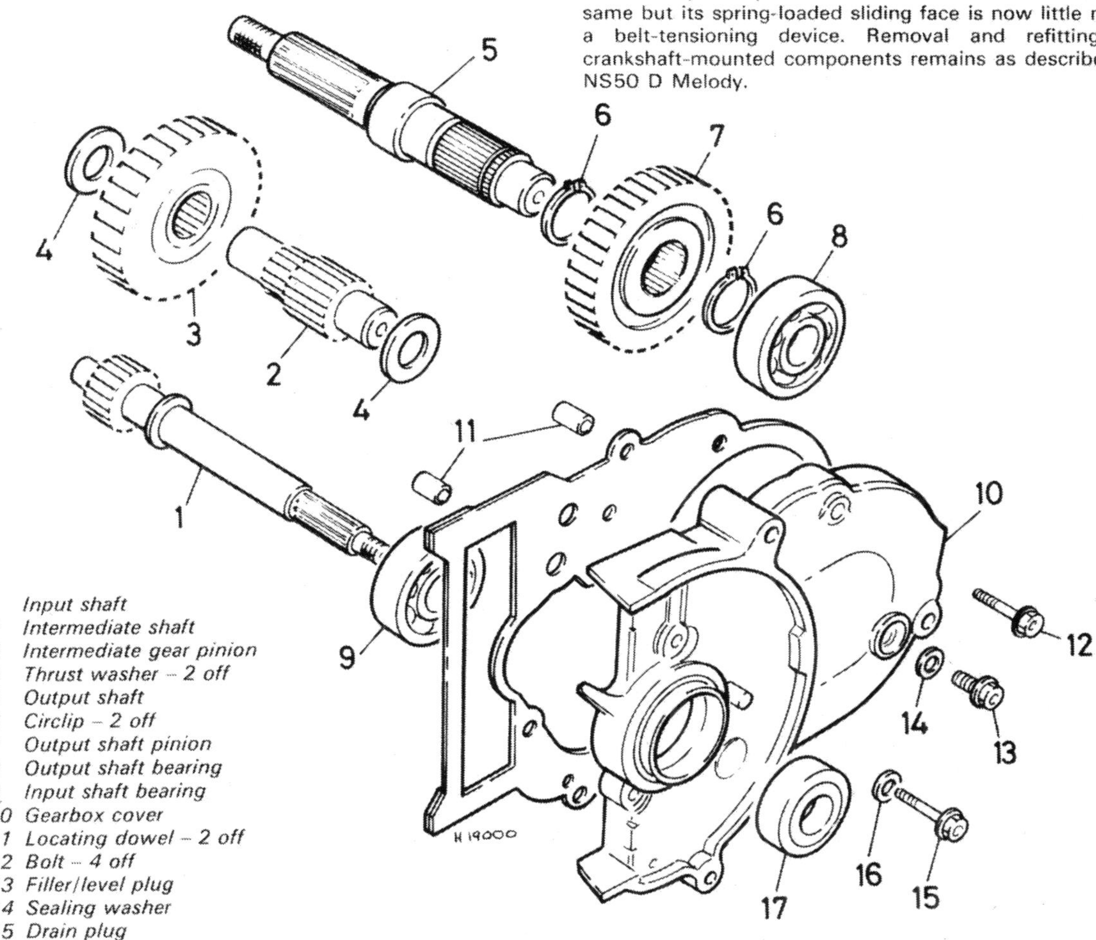

1 *Input shaft*
2 *Intermediate shaft*
3 *Intermediate gear pinion*
4 *Thrust washer – 2 off*
5 *Output shaft*
6 *Circlip – 2 off*
7 *Output shaft pinion*
8 *Output shaft bearing*
9 *Input shaft bearing*
10 *Gearbox cover*
11 *Locating dowel – 2 off*
12 *Bolt – 4 off*
13 *Filler/level plug*
14 *Sealing washer*
15 *Drain plug*
16 *Sealing washer*
17 *Oil seal*

Fig. 7.1 Reduction gearbox – NB50 model

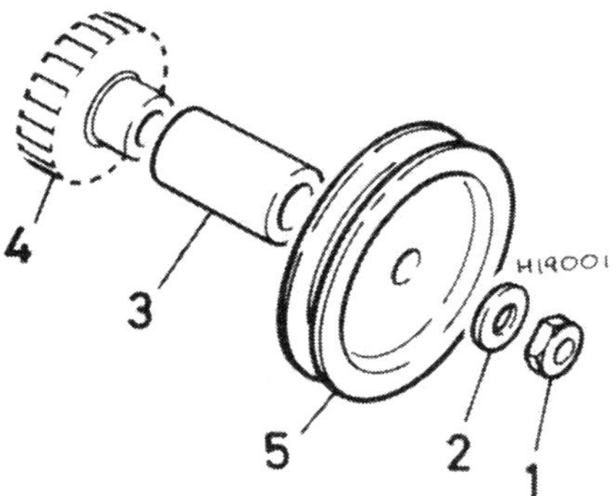

Fig. 7.2 Transmission drive pulley – NP50 model

1 Nut
2 Washer
3 Spacer
4 Kickstarter driven gear
5 Pulley

9 In the reduction gearbox, a thrust washer is fitted to the input shaft right-hand end.
10 Note that the engine mounting bracket is now a simpler tubular structure which no longer has the rubber bush attached to its centre. Removal and refitting procedures remain the same.

3 Removing the generator rotor: general – NB and NP50 models

1 Note that on these two later models it is no longer possible to use a legged puller as described in Chapter 1. There is no practical and safe alternative to the correct Honda service tools.
2 Hold the rotor using the holding tool, sold under part numbers 07725–0030000 or 07725–0010101, while its retaining nut is removed or refitted.
3 Remove the rotor using the extractor sold under part number 07733–0010000 or 07933-0110000. If the tool is of the centre bolt type, unscrew the centre bolt and screw the tool body **anticlockwise** as far as possible into the thread in the centre of the rotor (a *left-hand thread* is employed) until it seats firmly. Tighten the centre bolt down on to the crankshaft end and tap smartly on the bolt head with a hammer. The shock should jar the rotor free; if not, tighten the centre bolt further and tap again to release the rotor.

4 Cylinder barrel and piston: examination and renovation – ND50 model

1 For the ND50 M-C Melody II deluxe model it should be noted that oversized pistons and piston rings are listed. This means that, with reference to Chapter 1, Sections 22 and 23, the cylinder barrel can be rebored to the next oversize if it is ever found to be worn to the specified service limit or beyond.
2 Pistons and piston rings are available in oversizes of 0.25, 0.50, 0.75 and 1.00 mm (0.010, 0.020, 0.030 and 0.040 in). The amount of oversize will be stamped in the piston crown or etched on the piston ring top surface; do not forget to add the amount of oversize, if necessary, when determining the degree of wear that has taken place.
3 It is usual, when a rebore is required, to obtain the piston first so that the reborer can measure its diameter and bore the barrel to suit. After the rebore is complete, the bore should be honed lightly to provide a fine cross-hatched surface so that the new piston and rings can bed in correctly and the edges of the ports should be chamfered as described in Chapter 1, Section 22, paragraph 8.
4 Note that if a rebore is carried out, the existing piston and rings should be disregarded as they will be replaced by their oversized counterparts.

5 Fuel tank: removal and refitting – NB and NP50 models

While the work necessary to remove and refit the fuel tank is largely the same as the procedure given in Chapter 2, note that the battery retaining strap must be released as well as the four mounting bolts on NB50 models. On NP50 models the procedure is simplified by the fact that only the fuel feed pipe needs to be disconnected before the tank can be withdrawn; be careful to switch the tap to the 'Off' position before the feed pipe is disturbed.

6 Fuel tap: general – NP50 model

1 This model is fitted with a manually-operated three-position tap in the forward lower end of the fuel tank. Removal and refitting is as described in Chapter 2, but note that the tap is fitted with a filter gauze inside the tank.
2 If the tap becomes blocked at any time, try to clear it by removing it from the fuel tank and directing a blast of compressed air through it in the reverse direction to normal flow. If this fails to clear the blockage, or if the tap develops a fuel leak, it must be renewed; it is a sealed unit which cannot be dismantled for cleaning or repair.

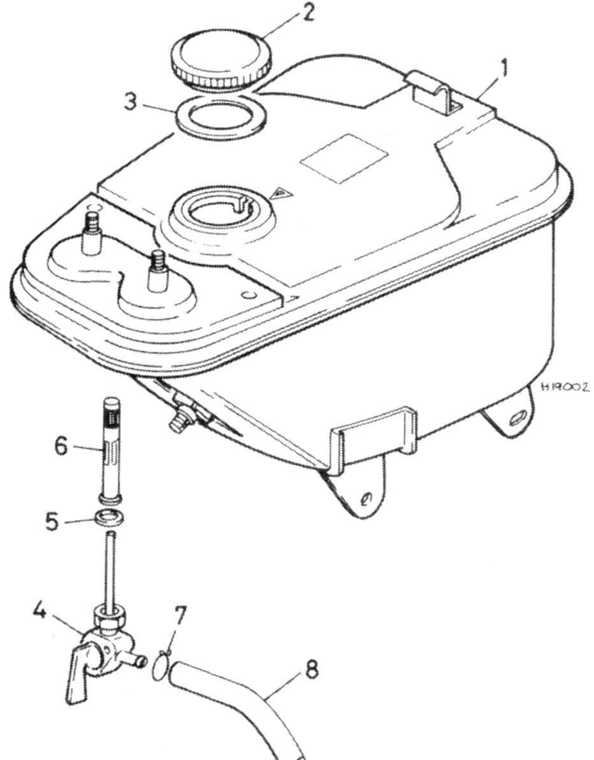

Fig. 7.3 Fuel tank and tap – NP50 model

1 Fuel tank
2 Filler cap
3 Gasket
4 Fuel tap
5 Sealing washer
6 Filter
7 Clip
8 Feed pipe

7 Carburettor: removal and refitting

1 Note that while the work necessary to remove and refit the carburettor is principally the same as that described in Chapter 2, there are some variations to be encountered on these later models. On ND50 and NB50 models the frame left-hand cover must be removed to gain access to the air filter; also on NB50 models the auto bystarter wires must be disconnected instead of the choke control pipes of the earlier models.

2 On NP50 models the choke plunger assembly must be withdrawn; apart from the throttle cable and fuel pipe there are no other components to be removed.

3 On all later models, note the presence of the heat insulating spacer between the carburettor and the intake stub; do not lose it and ensure that new gaskets or O-rings are used to seal the joint faces whenever it is disturbed.

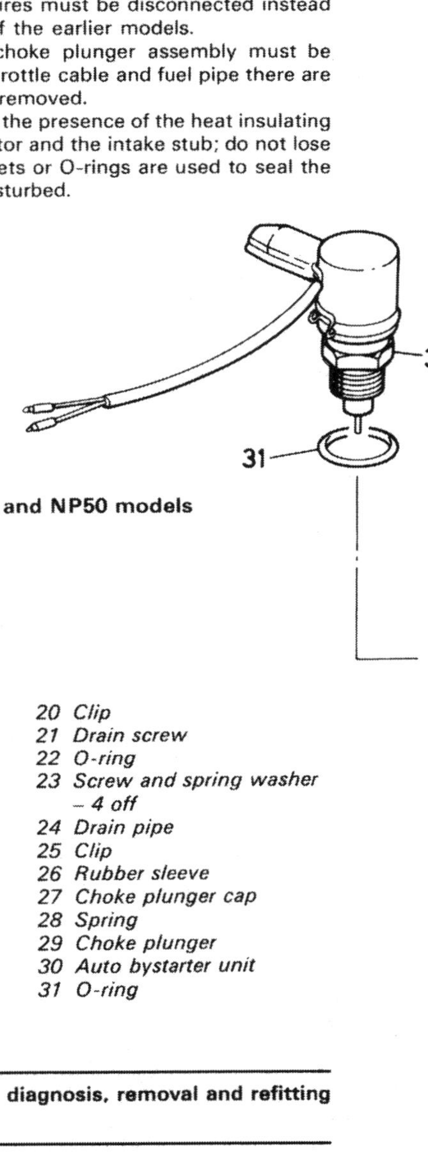

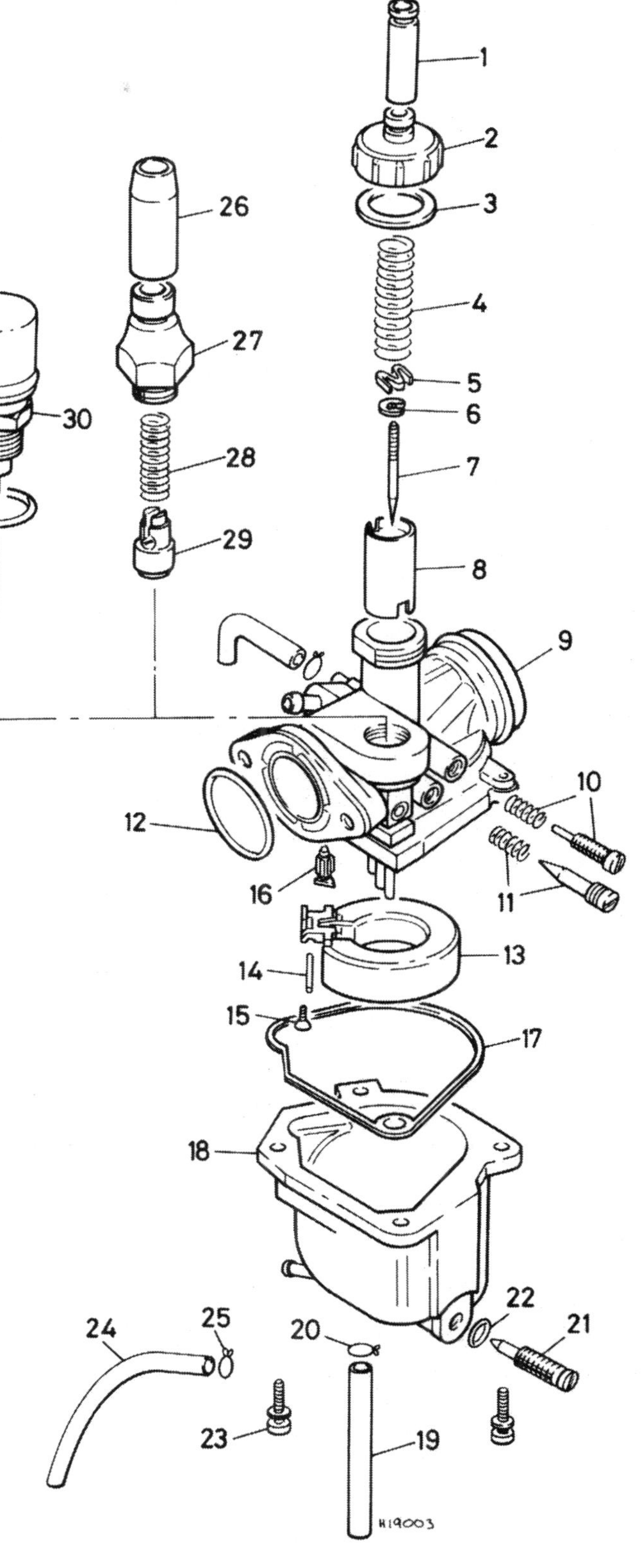

Fig. 7.4 Carburettor – NB and NP50 models

1 *Rubber sleeve*
2 *Carburettor top*
3 *Sealing washer*
4 *Return spring*
5 *Needle retaining spring clip*
6 *Needle position clip*
7 *Jet needle*
8 *Throttle valve*
9 *Carburettor body*
10 *Throttle stop screw*
11 *Pilot air screw*
12 *O-ring*
13 *Float*
14 *Float pivot pin*
15 *Screw*
16 *Float needle*
17 *Gasket*
18 *Float chamber*
19 *Overflow pipe*
20 *Clip*
21 *Drain screw*
22 *O-ring*
23 *Screw and spring washer – 4 off*
24 *Drain pipe*
25 *Clip*
26 *Rubber sleeve*
27 *Choke plunger cap*
28 *Spring*
29 *Choke plunger*
30 *Auto bystarter unit*
31 *O-ring*

8 Cold start device: fault diagnosis, removal and refitting – NB50 model

1 This model is equipped with a fully automatic cold start device, known by Honda as an auto bystarter. The unit consists of a conventional plunger-type valve operated by a small solenoid. It is housed in a black plastic casing mounted on top of the carburettor. When the ignition is first switched on, the plunger is raised, allowing a fuel-rich mixture to be fed to the engine to permit easy starting. After about five minutes of running, the valve is shut off, leaving the engine to run on its normal mixture.

2 The system is entirely unobtrusive in operation, and any fault will soon become apparent to the rider. If the valve fails to open when the engine is cold, starting will be difficult or even impossible, with a marked tendency towards stalling until the engine reaches full operating temperature. Conversely, if the

valve becomes stuck open, the engine will start normally, but will run irregularly when hot, with a noticeable increase in exhaust smoke.
3 If a fault is suspected, remove the left-hand side cover and trace back the leads from the unit to the connectors near the footboard. Disconnect the leads and measure the resistance between them using a multimeter. If the internal resistance is about 10 ohms, the unit is probably serviceable, but if infinite resistance is shown it can be assumed to be faulty and should be renewed. Note that the machine should have stood for at least 10 minutes before the test is made.
4 Further testing of the unit requires the removal of the carburettor and some method of applying compressed air to the richening circuit, via the air intake passage. This is located on the intake side of the carburettor throat, near the top. A foot or hand pump can be used, provided a suitable nozzle can be arranged to suit the passage.
5 Let the carburettor stand for at least 30 minutes, then apply air to the richening circuit. If it appears blocked, the cold start device is probably faulty. If air passes freely, connect the machine's battery to the cold start device leads and wait five minutes. If there is no resistance to the applied air pressure, renew the cold start device.
6 If the above checks are difficult to carry out at home, take the machine to a Honda dealer, who will be able to advise on the condition of the unit. Note that the sealed construction rules out any attempt at repair. The unit is held in place by two cross-head screws. Before refitting the cold start device, check the valve for signs of wear or damage, and renew the O-ring if it is damaged or broken.

9 Choke assembly: general – NP50 model

1 This model is fitted with a very simple manually-operated choke system which provides a rich mixture for starting. The handlebar lever, when applied, acts via a control cable on a plunger to open an auxiliary air passage in the carburettor body. The passage of air draws up sufficient fuel via the starter jet circuit to provide the enriched fuel/air mixture required. When the handlebar lever is returned, a spring fitted above the plunger forces it down on to its seat to close off the passage.
2 Although no cable adjustment is provided, it is essential to ensure that there is some free play in the cable at all times or the plunger may be raised off its seat and provide a permanently rich mixture. At regular intervals (see Routine Maintenance) the cable and lever should be dismantled, checked for wear or damage and lubricated.
3 Whenever the carburettor is removed, disconnect the choke cable by unscrewing the hexagon-shaped cap from the carburettor body and withdrawing the complete assembly.
4 Check the plunger carefully for signs of wear whenever the carburettor is overhauled. Check the plunger seat with special care, renewing it if it is found to be at all worn or scratched. The internal passages can be blown clear with compressed air, as described in Chapter 2.

10 Air filter: general

ND50 model

1 While the filter itself is exactly the same as that described in Routine Maintenance and in Chapter 2, the fuller bodywork of this model means that the frame left-hand cover must be removed first to gain access to the filter assembly.

NB and NP50 models

2 In the case of the NB50 model the frame left-hand cover must be removed first to gain access to the filter assembly. Both these models are fitted with a modified form of filter.
3 To release the element, remove the single retaining screw and withdraw the filter cover, followed by the metal supporting frame. The element can then be removed and cleaned as described in Routine Maintenance.

11 Oil tank: removal and refitting

1 The work necessary to remove and refit the oil tanks of the later models is outlined below. For full details refer to the relevant Sections of this Chapter and to Chapter 2, Section 17. Pay careful attention to the notes concerning refitting and check carefully for oil leaks.

ND50 models

2 Remove both frame covers and the battery. Remove its four mounting bolts and withdraw the rear luggage carrier, then unscrew the single mounting bolt and withdraw the battery carrier.
3 Remove the two remaining bolts which secure the tail lamp and flashing indicator lamp assembly and withdraw the unit until its wires can be disconnected. Remove the two mounting bolts and withdraw the tail lamp mounting bracket, followed by the rear mudguard. Remove the oil filler cap, pull clear the battery breather tube and withdraw the frame top rear cover.
4 Drain the oil into a clean container, disconnect the oil level switch wires, remove the single mounting bolt from the rear of the tank and lift the tank away.

NB50 model

5 Remove both frame covers and the battery, then unscrew the oil filler cap and withdraw the rubber drip tray around the filler neck. Remove the single bolt securing the battery case to the luggage carrier, unscrew the four carrier mounting bolts, and withdraw the carrier and battery case. Drain the oil into a clean container, disconnect the oil level switch wires, remove the single mounting bolt from the rear of the tank and lift the tank away.

NP50 model

6 Remove both frame covers and the battery, then disconnect the oil level switch wires and drain the oil into a clean container. Unscrew the two bolts which retain the battery carrier/oil tank support bracket, remove the bracket and withdraw the tank.

12 Cycle parts: modifications – NB and NP50 models

1 Apart from the modified components mentioned specifically in the relevant Sections of this Chapter, the following alterations should be noted:
2 NP50 models are fitted with telescopic forks and very different handlebar mountings; refer to the following Sections of this Chapter for details.
3 Note that the front mudguard on NP50 models is retained by two bolts.
4 On both NB50 and NP50 models the rear suspension unit is fitted with a nylon spring guide and a separate spring seat, these being fitted between the spring and the unit body.

13 Handlebar: removal and refitting – NP50 model

1 Working as described in Section 17 dismantle the handlebar nacelle and remove the front body panel, then disconnect the ignition switch lead at its block connector and remove the nacelle top cover.
2 Disconnect the flashing indicator lamp wires and both handlebar switch wires. There is no need to remove either lamp assembly from the handlebar, unless required.
3 Remove the choke lever pivot screw and withdraw the lever, then disconnect the cable. Remove the two screws

clamping both halves of each switch and withdraw the handlebar switches. Unscrew the handlebar clamp bolt and tap the handlebar sharply upwards to release it.

4 On reassembly push the handlebar down on to the steering stem then rotate it and adjust its height until the clamp bolt can be refitted; it must engage with the groove cut in the steering stem to locate the handlebar and provide additional security by acting as a cotter pin. Tighten the bolt to a torque setting of 43 – 65 lbf ft (6.0 – 9.0 kgf m).

5 The remainder of the reassembly procedure is a reversal of the removal sequence but note that the handlebar bearing surface should be smeared with a light film of grease before the twistgrip is refitted. Note also that the joint faces at the rear of each switch must align with the punch marks in the handlebar to ensure that they are correctly positioned.

6 Ensure that all disturbed electrical connections are correctly made and that all components function correctly when work is complete. Check also that all controls are working smoothly and correctly throughout their full range before starting the engine and taking the machine out on the road.

14 Front fork and steering head assembly: removal – NP50 model

1 The procedure laid out in Chapter 4 is basically correct for these models provided that attention is given to the relevant modified procedures laid out in this Chapter.

2 The front forks can be removed as a complete unit, by following the full procedure, or they can be dismantled whilst in place on the machine if only the suspension components are to

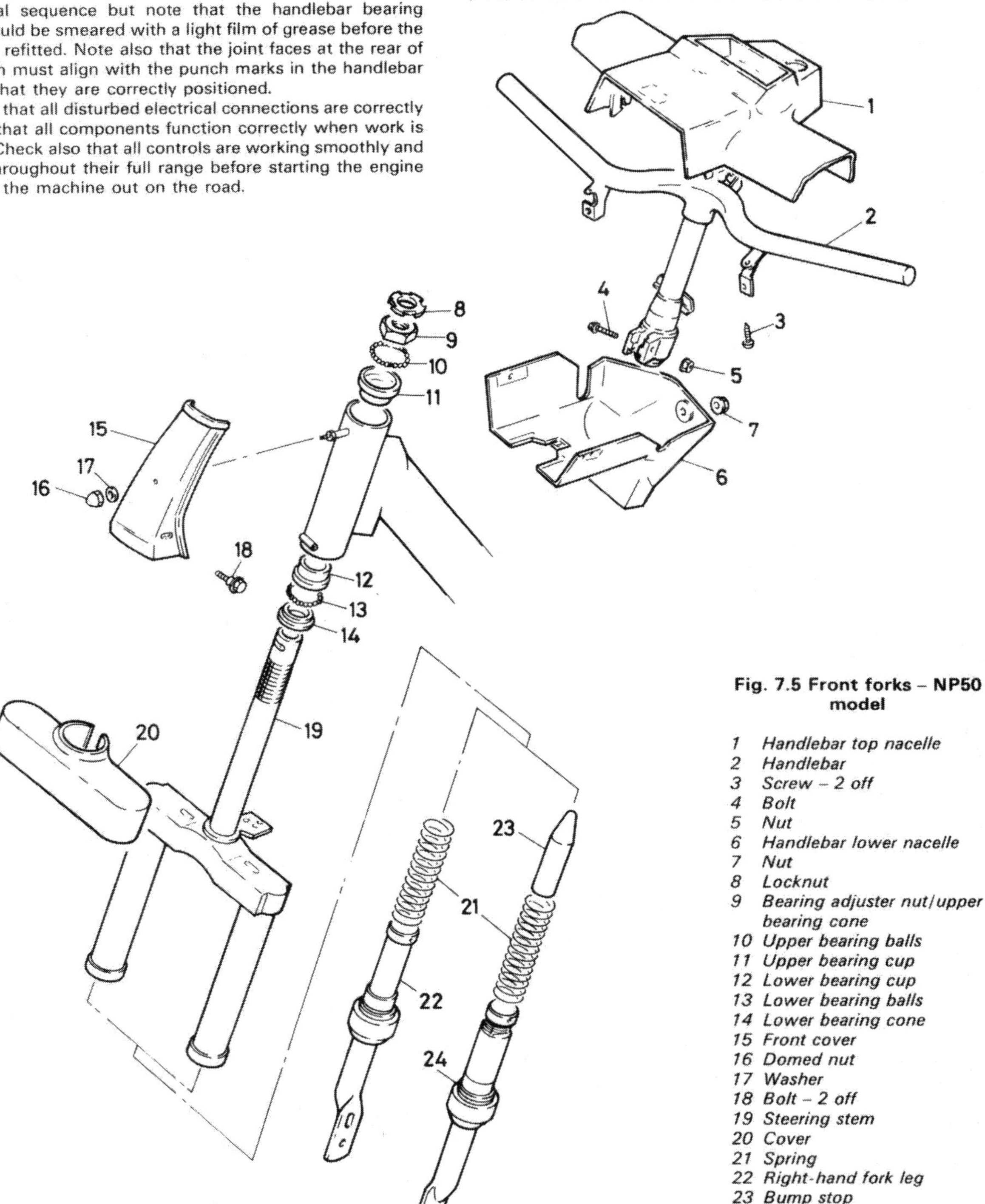

Fig. 7.5 Front forks – NP50 model

1 Handlebar top nacelle
2 Handlebar
3 Screw – 2 off
4 Bolt
5 Nut
6 Handlebar lower nacelle
7 Nut
8 Locknut
9 Bearing adjuster nut/upper bearing cone
10 Upper bearing balls
11 Upper bearing cup
12 Lower bearing cup
13 Lower bearing balls
14 Lower bearing cone
15 Front cover
16 Domed nut
17 Washer
18 Bolt – 2 off
19 Steering stem
20 Cover
21 Spring
22 Right-hand fork leg
23 Bump stop
24 Left-hand fork leg

be overhauled. To carry out the work without removing the complete assembly from the machine, proceed as follows:

3 Support the machine on its centre stand on level ground so that it cannot tip forwards and remove the front wheel.

4 Slide down the dust cover at the bottom of each steering stem fixed leg, then use circlip pliers to withdraw the retaining circlip from inside the bottom of each fixed leg. Pull the plated sliding legs downwards with a sharp tug to release them. Check that all components, including the spring and bump stop, have been removed.

5 Do not attempt to dismantle the sliding leg assemblies any further; no spare parts are available with which they can be reconditioned. Be very careful to keep separate the components of the two legs. If part worn items are mixed on reassembly, the rate of wear will be greatly accelerated.

15 Front fork: examination and renovation – NP50 model

1 Start by thoroughly cleaning all components, not forgetting the inside of the steering stem fixed legs. Remove all traces of old grease, dirt and corrosion.

2 Check each component for obvious signs of wear or damage and renew any item that is too worn or damaged to be of further use. Refer to the relevant Sections of Chapter 4 for information on checking the steering head bearings and other components not specifically mentioned here.

3 As previously mentioned, the sliding leg assemblies are sealed and must be renewed if either is found to be at all worn or damaged. Check carefully the bearing surfaces of the top (fixed) and bottom (sliding) bushes looking for scratches or other signs of excessive wear; do not forget to check the matching bearing surfaces inside the fixed fork legs.

4 To check for wear, insert each sliding leg into its respective location and feel for free play at all points throughout its full travel. Since no service limits are specified, it is largely a matter of personal experience to decide how much, if any, free play is permissible before renewal is required. If in any doubt, take the complete assembly to an authorised Honda dealer for an expert opinion.

5 If any undue stiffness is encountered during this check, the offending leg should be checked carefully to ensure that it is not bent. If severe warpage (ie more than 0.2 mm/0.008 in) is discovered, the damaged components must be renewed. Attempts at straightening are not recommended as the forks are too light in construction; the risk of subsequent fatigue failure is very real. Furthermore it would be difficult to straighten successfully such irregularly-shaped components.

16 Front fork: refitting – NP50 model

1 Smear a liberal coat of grease over the springs, sliding leg assemblies and inside the fixed fork legs. Fit the bump stop, spring and sliding leg together as a single unit and insert each assembly into its original fixed leg.

2 Press the bottom bush upwards into place until the retaining circlip can be refitted in its groove. Check that each leg moves smoothly and correctly throughout its travel, then refit the dust seals and the front wheel.

3 Check that all disturbed components have been correctly refitted and that all nuts and bolts are securely fastened. Before taking the machine out on the road, check the operation of the front suspension and that all controls are correctly adjusted and working properly.

17 Body panels: removal and refitting

Handlebar nacelle – ND50 model

1 To dismantle the handlebar nacelle on ND50 models, proceed as described in Chapter 4, Section 7, but note that the speedometer assembly is held by two screws to the nacelle upper half.

Handlebar nacelle – NB50 model

2 Remove from the underside of the handlebar the screw securing each flashing indicator lamp assembly, disconnect the wires, and withdraw both lamps. Remove the single retaining screw from the headlamp bottom edge, withdraw the headlamp unit, and disconnect its wires.

3 Unscrew the two nuts securing the top cover to the handlebar then remove the three screws from the rear of the nacelle which secure the top and bottom halves of the assembly. Disconnect the speedometer drive cable and all electrical components, and lift the nacelle away.

Handlebar nacelle – NP50 model

4 Remove the single retaining screw from the headlamp bottom edge, withdraw the headlamp unit, and disconnect its wires. Unscrew the single cap nut from the rear of the nacelle. Working from the front of the machine very carefully, disengage the two clips, one on each side of the headlamp opening, which secure the nacelle top and bottom covers.

5 The top cover is retained to the handlebar by two screws, one on each end behind the indicator lamp assemblies. Remove these screws and lift the top cover until the speedometer drive cable can be unscrewed and all electrical components disconnected. The nacelle top half can then be lifted away; the speedometer is secured to it by two screws.

6 The nacelle bottom half cannot be removed until the handlebars are withdrawn.

Front body panel and footrest panel – all models

7 In the main, the instructions given in Section 8 of Chapter 4 apply equally when removing or refitting these components on the later models. Front baskets (where fitted) should be removed first; these are retained by three cap nuts.

8 Storage pockets fitted on the inside of the front panel are retained by clips on early models and by screws on later models; the mounting points are to be found at each corner to secure it to the front panel and/or the footrest panel.

9 On ND50 and NB50 models the steering column cover is retained by a single screw; once this is withdrawn the front panel's remaining mountings can be seen easily.

10 On NP50 models remove the two bolts and the cap nut which secure the steering column front cover, withdraw the cover, and unscrew the single bolt which secures the front body panel to the steering column. Remove the panel.

Rear frame covers – ND and NB50 models

11 Raise the seat and open the storage compartment. First unscrew the cap nuts or cap nut and bolt which clamp both covers at their front mating surface, then unscrew the two bolts inside the storage compartment which secure the right-hand cover to the frame.

12 Unscrew the single bolt securing the rear end of each cover to the tail lamp or to its mounting bracket. Lift away the covers.

Rear frame covers – NP50 model

13 Raise the seat. Unscrew the bolt and the cap nut which clamp both covers at their front mating surface, then unscrew the two bolts and the cap nut which retain the rear luggage carrier. Withdraw the carrier and lift away the covers.

Bodywork – all models

14 When removing and refitting plastic components, take great care not to force them into or out of position; the panels should fit together easily and neatly. If any undue hindrance is

encountered, stop immediately before damage is done and check carefully that all fasteners have been removed and that there are no clips or similar still holding the component in place.
15 Before starting work, be careful to examine all components, checking their mountings and fit, and devise a logical sequence of operations. For example, it may well prove very difficult to remove the footboard without removing the rear covers first.

18 Front wheel: removal and refitting

ND50 model

1 Before the wheel can be withdrawn, the plated cover must be removed from each of the front fork links; each cover is retained by a single bolt.

NB50 model

2 Before the wheel can be withdrawn, the black plastic covers, each retained by one bolt, must be removed from the front fork links. Note also that the front brake cable is now disconnected by unscrewing the adjuster nut and withdrawing the cable first from its abutment on the brake backplate then from the operating arm trunnion.

NP50 model

3 Although the forks are very different in appearance, the front wheel removal and refitting procedure remains as described in Chapter 5. The only significant difference is that the front brake cable is now disconnected by unscrewing the adjuster nut and withdrawing the cable first from its abutment on the brake backplate, then from the operating arm trunnion.

19 Rear wheel: removal and refitting

While the procedure remains largely similar to that described in Chapter 5, refer to Section 17 of this Chapter for details of the work necessary to gain access to the exhaust system. Note also on the NB50 and NP50 models that the rear wheel is retained by a flanged nut only, no separate washer being fitted.

20 Front brake: general – NB and NP50 models

1 The front brake operating arm is fitted with an external return spring in addition to the main return spring described in Chapter 5.
2 The brake adjuster mechanism has been revised so that the adjusting nut is now screwed on to a thread at the cable lower end. The adjusting procedure remains as described in Routine Maintenance.

21 Charging/lighting coil: testing – NB and ND50 models

To test the charging/lighting coil on these two later models, proceed as described in Chapter 6.5 but note the different results to be expected from a sound coil:

White wire to earth .... 0.2 – 2.0 ohm
Yellow wire to earth .... 0.1 – 1.0 ohm

22 Regulator/rectifier unit: location and testing – NB and ND50 models

1 These models are fitted with a regulator/rectifier unit which combines the function of the silicon rectifier (Chapter 6, Section 5) with a degree of voltage control.
2 The unit is a finned, sealed metal device mounted underneath the frame rear covers on the left-hand side of the machine. It is to be found just beneath the ignition HT coil on ND50 models and to the immediate rear of the suspension unit top mounting on NB50 models. To remove the unit, unscrew its single mounting bolt and disconnect the multi-pin connector to release the wiring.
3 The unit requires no maintenance other than a regular check to ensure that it is clean and dry and that its connections and mounting point are clean and securely fastened.
4 To test the unit, disconnect its wires and use a good quality multimeter set to the appropriate resistance scale to measure the resistance between the various pairs of terminals as shown in the accompanying illustration, comparing the readings obtained with those specified in the accompanying test chart. Note that the chart is drawn up for a positive earth meter; the results will be reversed if a negative earth meter is used.
5 If any one reading is significantly different from that specified, the unit can be assumed to be faulty. However, always have your findings confirmed by an authorised Honda dealer using the correct service tools before renewing the unit.

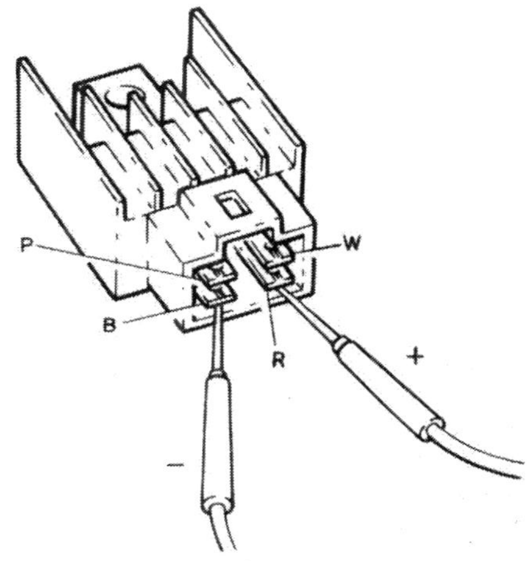

− \ +	RED	BLACK	PINK	WHITE
RED		∞	∞	∞
BLACK	∞		1 - 5kΩ	∞
PINK	∞	1 - 5kΩ		∞
WHITE	0.5 - 10kΩ	∞	∞	

H19005

Fig. 7.6 Regulator/rectifier unit test – NB and ND50 models

23 Resistors: general

1 With reference to Chapter 6, Section 10 for details of test procedures, note the following modifications:

ND50 model

2 This model is fitted with two resistors, both of which are identical to that described in Chapter 6 and are to be found in the same place. The unit which is connected to the Pink/White

wire from the wiring loom is connected only to the lighting switch and soaks up excess current generated when the engine is running without the headlamp in use.
3 The remaining resistor unit, which is connected to the Pink wire from the wiring loom, provides a form of voltage control. It is connected both to the regulator side of the regulator/rectifier unit and to the lighting switch so that when the lights are switched off, the resistor is shorted to earth, which results in a drop in the regulator output.
4 If persistent bulb-blowing problems or other symptoms of over-charging are experienced which indicate a fault in either resistor, both can be tested as described in Chapter 6.

NB50 model

5 This model is also fitted with two resistors; their function is exactly as described above for the ND50 model and the test procedure is as described in Chapter 6. One of the units is to be found bolted to the left-hand side of the frame rear section, immediately under the oil tank, while the second is bolted to the oil tank rear mounting, just in front of the ignition HT coil.

NP50 model

6 This model is fitted with a less sophisticated electrical system in which the two resistors provide a large measure of charge control. The main unit, resistor 'A', has a resistance value of 3 ohms and is connected to the Pink wire leading from the main wiring loom. It is connected to the ignition switch to soak up all surplus current generated when the engine is running with lights switched off.
7 The second unit, resistor 'B', has a value of 1.8 ohms and is connected via the Pink/White wire to the lighting switch to soak up the excess current present when the engine is running and the lights are switched on, to the 'PO' position.
8 Both units are to be found bolted to the underside of the right-hand footboard supports, resistor 'A' being nearest the engine unit. Test procedures are as described in Chapter 6, but note the lesser values given above for each of these units.

24 Oil level warning circuit: testing – NB and NP50 models

NB50 model

1 This model uses a greatly simplified circuit to provide a warning that the level of oil in the tank is dangerously low. The oil level sensor consists of a rubber-encased float switch mounted in the top of the oil tank. It is a simple and robust unit, and is unlikely to give trouble. To gain access to the sensor for testing a considerable amount of preliminary dismantling is necessary to reach the top of the oil tank as described in Section 17. Unplug the sensor leads from the top of the unit, and pull it out of the oil tank.
2 Using a multimeter or a battery and bulb continuity tester, check that the sensor conducts when the float is in the fully down position, and is isolated when the float is raised. If all seems well, reconnect the sensor leads and switch on the ignition. Try raising and lowering the float and note the effect on the warning lamp. If the problem persists, check the wiring for breaks or loose connections.
3 The actual warning lamp is an LED (Light Emitting Diode) assembly about which there is no information; it must, therefore, be assured that the complete unit must be renewed if a fault is suspected. Care must be taken to eliminate all other possible causes of failure before this is done.

NP50 model

4 The circuit on this model consists of a switch, which is pressed into the top of the oil tank, and an LED unit which is pressed into the base of the speedometer. There is no information available on test procedures for either unit; if a fault develops, therefore, all that can be done is to check the power supply and wiring. If the fault persists the switch and LED unit will have to be tested by the substitution of new components until the fault is located. When checking the wiring note that a 500 ohm, 1 watt resistor is fitted into the Green/Red wire between the switch and LED.

25 Headlamp: bulb renewal – NB and NP50 models

Note that the headlamp rim and reflector units on these models are retained by the screw set in the unit lower edge. This also serves as a beam height adjuster screw and so the beam must always be checked whenever it has been disturbed to ensure that it complies with local legislation.

26 Flashing indicator lamp; bulb renewal – NB50 model

Although the lamp assemblies fitted to this model are very different in appearance, the working procedures are unchanged. The front lamp lenses are each retained by a single screw at their outboard ends and by a clip at their inner ends. The rear lamp lens is retained by two screws and must be removed to gain access to any of the three bulbs; all bulbs are of the bayonet type, as described in Chapter 6.

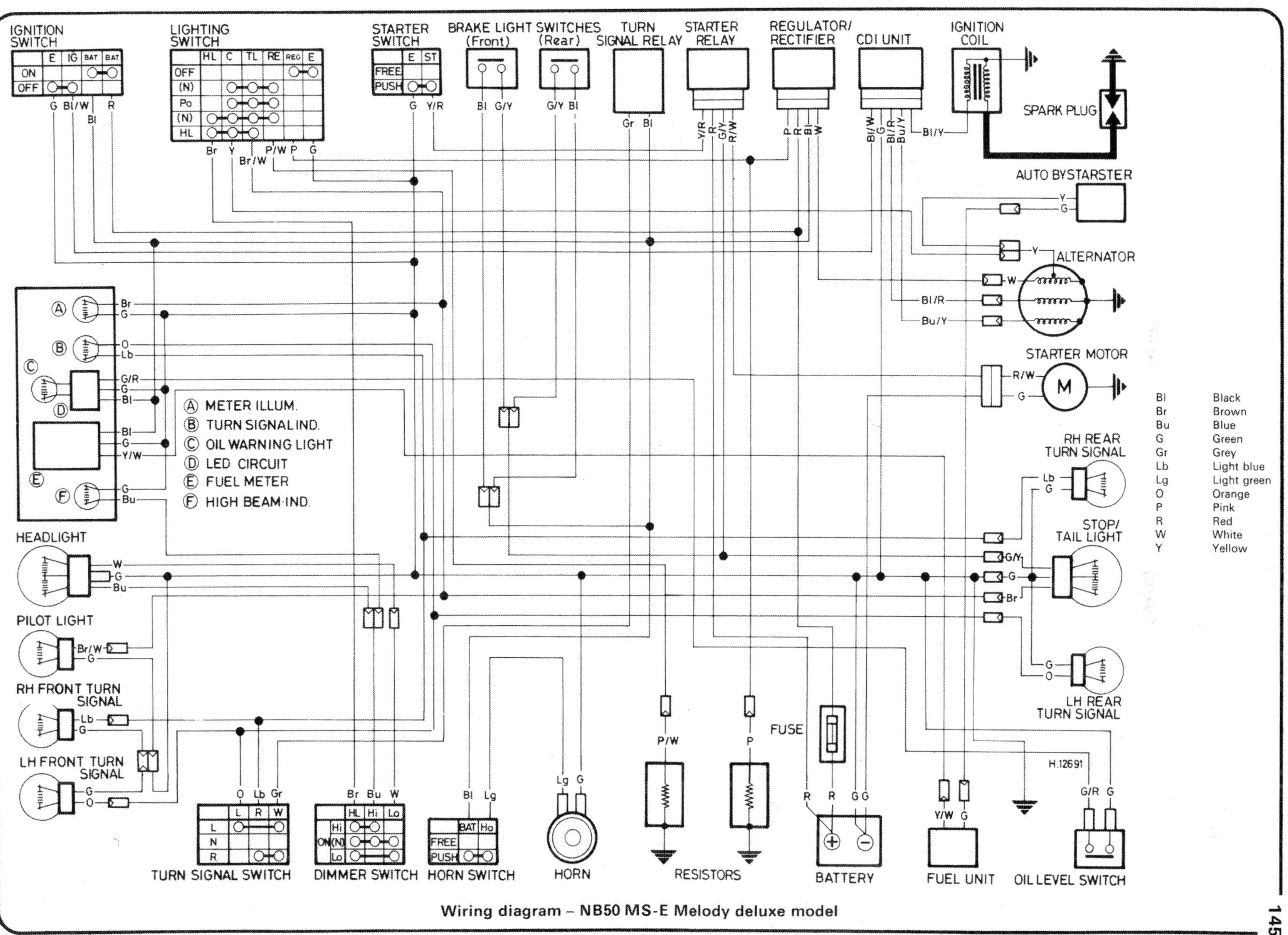

Wiring diagram – NB50 MS-E Melody deluxe model

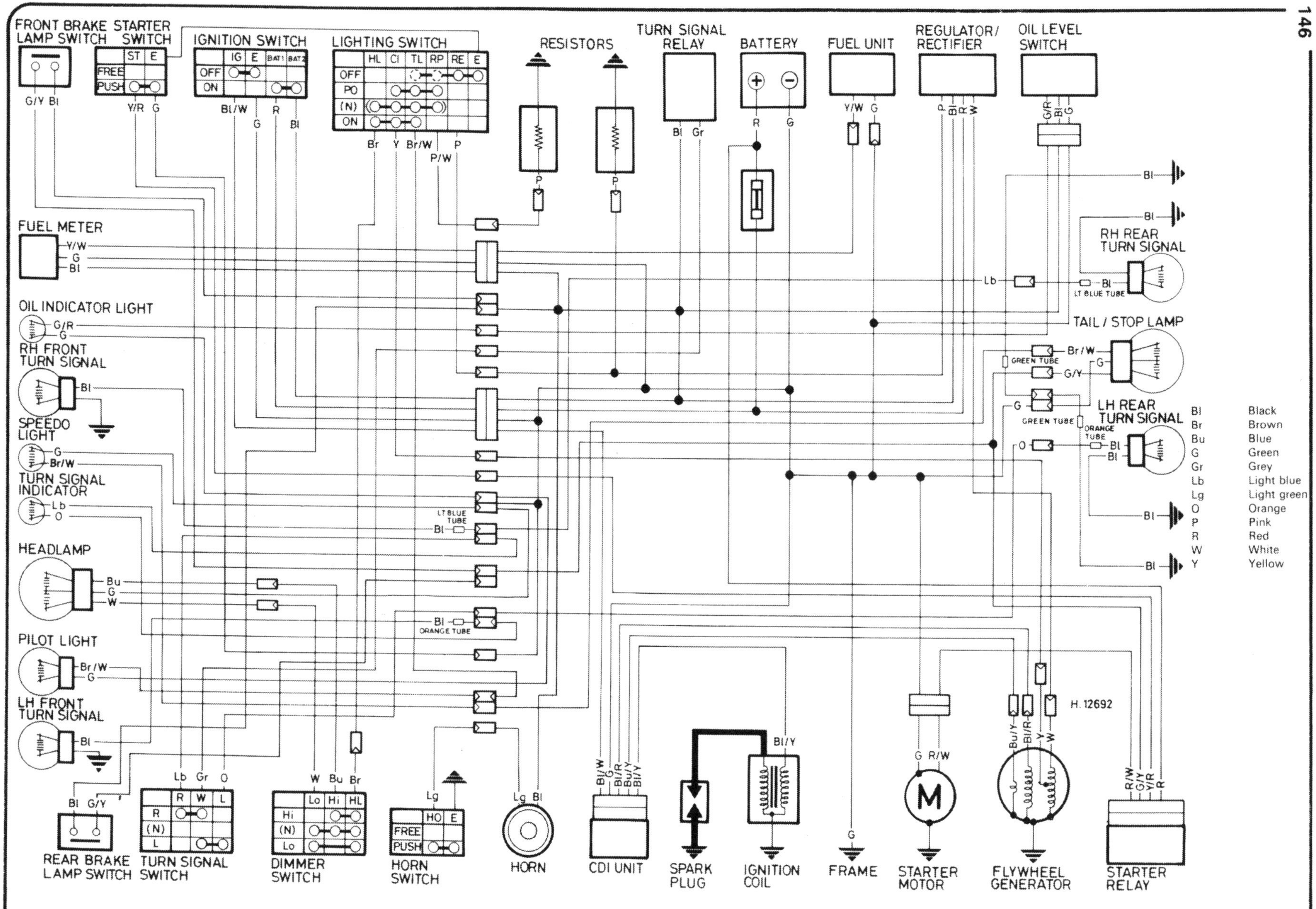

FRONT BRAKE LAMP SWITCH

STARTER SWITCH

IGNITION SWITCH

LIGHTING SWITCH

RESISTORS

TURN SIGNAL RELAY

BATTERY

FUEL UNIT

REGULATOR/ RECTIFIER

OIL LEVEL SWITCH

FUEL METER

OIL INDICATOR LIGHT

RH FRONT TURN SIGNAL

SPEEDO LIGHT

TURN SIGNAL INDICATOR

HEADLAMP

PILOT LIGHT

LH FRONT TURN SIGNAL

RH REAR TURN SIGNAL

TAIL / STOP LAMP

LH REAR TURN SIGNAL

LT BLUE TUBE

GREEN TUBE

ORANGE TUBE

REAR BRAKE LAMP SWITCH

TURN SIGNAL SWITCH

DIMMER SWITCH

HORN SWITCH

HORN

CDI UNIT

SPARK PLUG

IGNITION COIL

FRAME

STARTER MOTOR

FLYWHEEL GENERATOR

STARTER RELAY

H. 12692

Bl	Black
Br	Brown
Bu	Blue
G	Green
Gr	Grey
Lb	Light blue
Lg	Light green
O	Orange
P	Pink
R	Red
W	White
Y	Yellow

Wiring diagram – ND50 M-C Melody II deluxe model

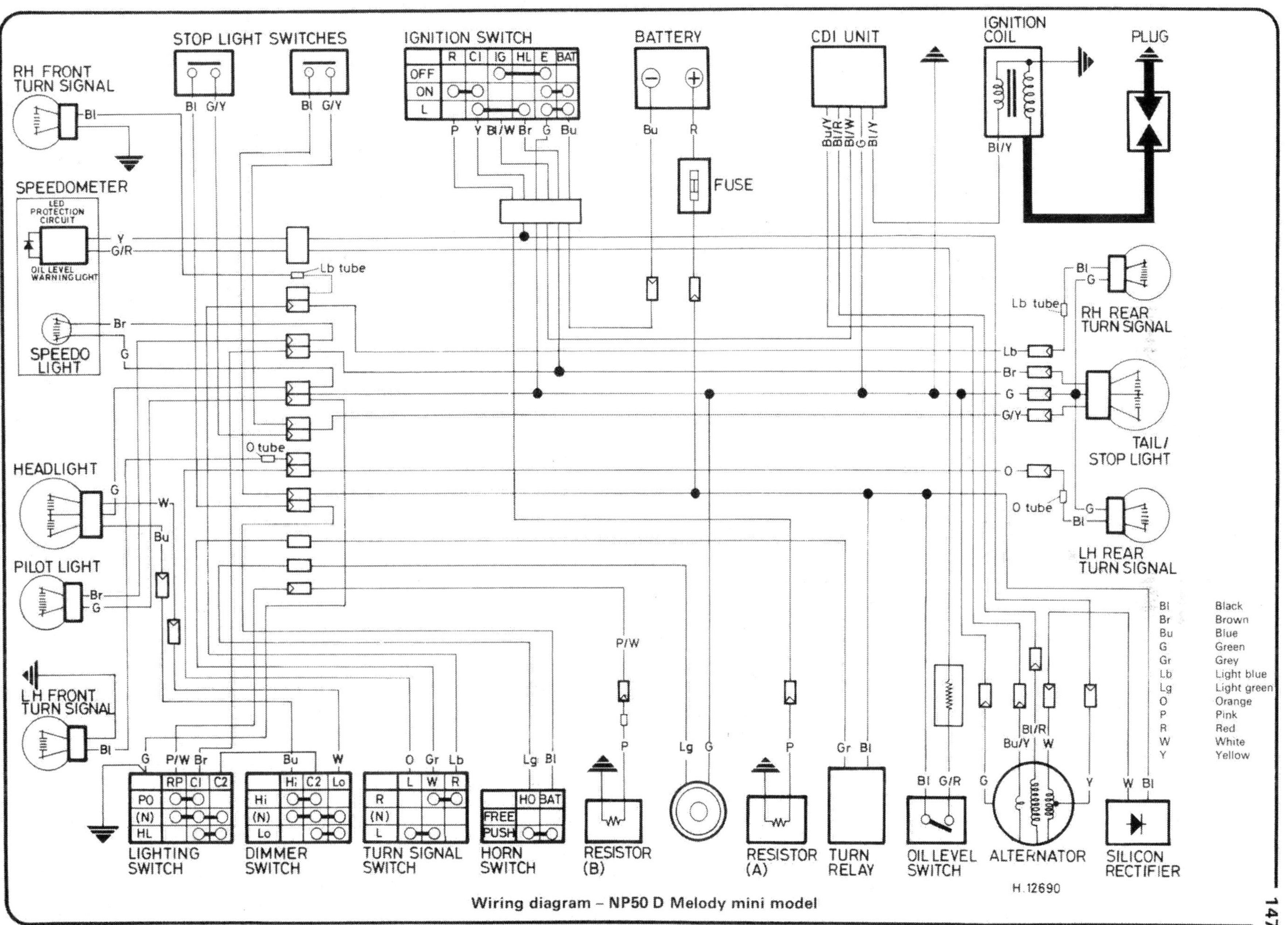

Wiring diagram – NP50 D Melody mini model

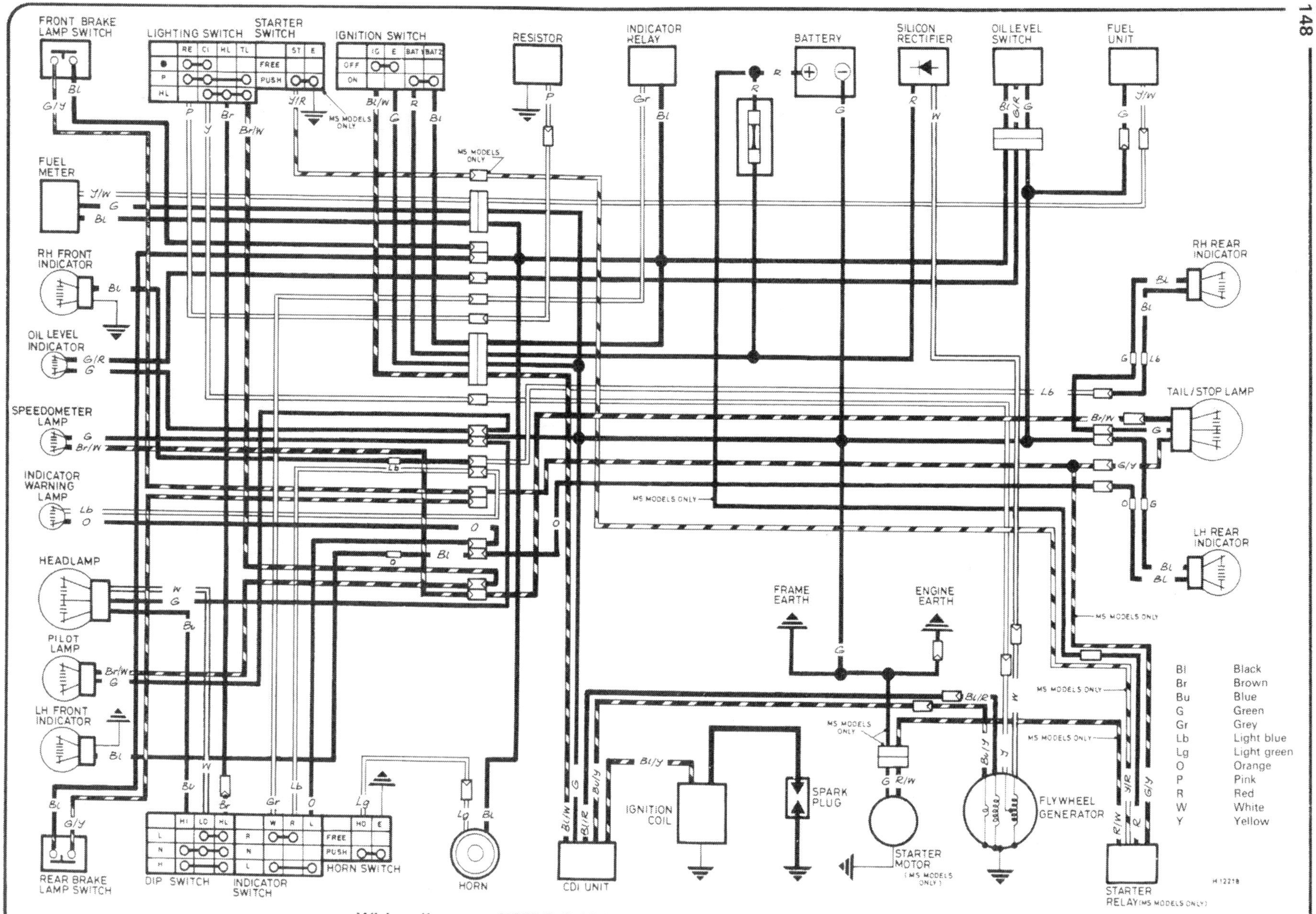

FRONT BRAKE LAMP SWITCH
LIGHTING SWITCH
STARTER SWITCH
IGNITION SWITCH
RESISTOR
INDICATOR RELAY
BATTERY
SILICON RECTIFIER
OIL LEVEL SWITCH
FUEL UNIT
FUEL METER
RH FRONT INDICATOR
OIL LEVEL INDICATOR
SPEEDOMETER LAMP
INDICATOR WARNING LAMP
HEADLAMP
PILOT LAMP
LH FRONT INDICATOR
REAR BRAKE LAMP SWITCH
DIP SWITCH
INDICATOR SWITCH
HORN SWITCH
HORN
CDI UNIT
IGNITION COIL
SPARK PLUG
STARTER MOTOR (MS MODELS ONLY)
FLYWHEEL GENERATOR
STARTER RELAY (MS MODELS ONLY)
FRAME EARTH
ENGINE EARTH
RH REAR INDICATOR
TAIL/STOP LAMP
LH REAR INDICATOR
MS MODELS ONLY

Bl	Black
Br	Brown
Bu	Blue
G	Green
Gr	Grey
Lb	Light blue
Lg	Light green
O	Orange
P	Pink
R	Red
W	White
Y	Yellow

H12218

Wiring diagram – NS50 D-B Melody and NS50 MS-B Melody deluxe models

Conversion factors

Length (distance)						
Inches (in)	X	25.4	= Millimetres (mm)	X	0.0394	= Inches (in)
Feet (ft)	X	0.305	= Metres (m)	X	3.281	= Feet (ft)
Miles	X	1.609	= Kilometres (km)	X	0.621	= Miles
Volume (capacity)						
Cubic inches (cu in; in^3)	X	16.387	= Cubic centimetres (cc; cm^3)	X	0.061	= Cubic inches (cu in; in^3)
Imperial pints (Imp pt)	X	0.568	= Litres (l)	X	1.76	= Imperial pints (Imp pt)
Imperial quarts (Imp qt)	X	1.137	= Litres (l)	X	0.88	= Imperial quarts (Imp qt)
Imperial quarts (Imp qt)	X	1.201	= US quarts (US qt)	X	0.833	= Imperial quarts (Imp qt)
US quarts (US qt)	X	0.946	= Litres (l)	X	1.057	= US quarts (US qt)
Imperial gallons (Imp gal)	X	4.546	= Litres (l)	X	0.22	= Imperial gallons (Imp gal)
Imperial gallons (Imp gal)	X	1.201	= US gallons (US gal)	X	0.833	= Imperial gallons (Imp gal)
US gallons (US gal)	X	3.785	= Litres (l)	X	0.264	= US gallons (US gal)
Mass (weight)						
Ounces (oz)	X	28.35	= Grams (g)	X	0.035	= Ounces (oz)
Pounds (lb)	X	0.454	= Kilograms (kg)	X	2.205	= Pounds (lb)
Force						
Ounces-force (ozf; oz)	X	0.278	= Newtons (N)	X	3.6	= Ounces-force (ozf; oz)
Pounds-force (lbf; lb)	X	4.448	= Newtons (N)	X	0.225	= Pounds-force (lbf; lb)
Newtons (N)	X	0.1	= Kilograms-force (kgf; kg)	X	9.81	= Newtons (N)
Pressure						
Pounds-force per square inch (psi; lbf/in^2; lb/in^2)	X	0.070	= Kilograms-force per square centimetre (kgf/cm^2; kg/cm^2)	X	14.223	= Pounds-force per square inch (psi; lbf/in^2; lb/in^2)
Pounds-force per square inch (psi; lbf/in^2; lb/in^2)	X	0.068	= Atmospheres (atm)	X	14.696	= Pounds-force per square inch (psi; lbf/in^2; lb/in^2)
Pounds-force per square inch (psi; lbf/in^2; lb/in^2)	X	0.069	= Bars	X	14.5	= Pounds-force per square inch (psi; lbf/in^2; lb/in^2)
Pounds-force per square inch (psi; lbf/in^2; lb/in^2)	X	6.895	= Kilopascals (kPa)	X	0.145	= Pounds-force per square inch (psi; lbf/in^2; lb/in^2)
Kilopascals (kPa)	X	0.01	= Kilograms-force per square centimetre (kgf/cm^2; kg/cm^2)	X	98.1	= Kilopascals (kPa)
Millibar (mbar)	X	100	= Pascals (Pa)	X	0.01	= Millibar (mbar)
Millibar (mbar)	X	0.0145	= Pounds-force per square inch (psi; lbf/in^2, lb/in^2)	X	68.947	= Millibar (mbar)
Millibar (mbar)	X	0.75	= Millimetres of mercury (mmHg)	X	1.333	= Millibar (mbar)
Millibar (mbar)	X	1.40	= Inches of water (inH_2O)	X	0.714	= Millibar (mbar)
Millimetres of mercury (mmHg)	X	1.868	= Inches of water (inH_2O)	X	0.535	= Millimetres of mercury (mmHg)
Inches of water (inH_2O)	X	27.68	= Pounds-force per square inch (psi, lbf/in^2, lb/in^2)	X	0.036	= Inches of water (inH_2O)
Torque (moment of force)						
Pounds-force inches (lbf in; lb in)	X	1.152	= Kilograms-force centimetre (kgf cm; kg cm)	X	0.868	= Pounds-force inches (lbf in; lb in)
Pounds-force inches (lbf in; lb in)	X	0.113	= Newton metres (Nm)	X	8.85	= Pounds-force inches (lbf in; lb in)
Pounds-force inches (lbf in; lb in)	X	0.083	= Pounds-force feet (lbf ft; lb ft)	X	12	= Pounds-force inches (lbf in; lb in)
Pounds-force feet (lbf ft; lb ft)	X	0.138	= Kilograms-force metres (kgf m; kg m)	X	7.233	= Pounds-force feet (lbf ft; lb ft)
Pounds-force feet (lbf ft; lb ft)	X	1.356	= Newton metres (Nm)	X	0.738	= Pounds-force feet (lbf ft; lb ft)
Newton metres (Nm)	X	0.102	= Kilograms-force metres (kgf m; kg m)	X	9.804	= Newton metres (Nm)
Power						
Horsepower (hp)	X	745.7	= Watts (W)	X	0.0013	= Horsepower (hp)
Velocity (speed)						
Miles per hour (miles/hr; mph)	X	1.609	= Kilometres per hour (km/hr; kph)	X	0.621	= Miles per hour (miles/hr; mph)
Fuel consumption*						
Miles per gallon, Imperial (mpg)	X	0.354	= Kilometres per litre (km/l)	X	2.825	= Miles per gallon, Imperial (mpg)
Miles per gallon, US (mpg)	X	0.425	= Kilometres per litre (km/l)	X	2.352	= Miles per gallon, US (mpg)

Temperature

Degrees Fahrenheit = (°C x 1.8) + 32

Degrees Celsius (Degrees Centigrade; °C) = (°F - 32) x 0.56

**It is common practice to convert from miles per gallon (mpg) to litres/100 kilometres (l/100km), where mpg (Imperial) x l/100 km = 282 and mpg (US) x l/100 km = 235*

Index

D

E

F

Zeitfracht Medien GmbH
Ferdinand-Jühlke-Straße 7
99095 Erfurt, Deutschland
produktsicherheit@kolibri360.de